"If we're serious about really addressing climate change, we need to become energy and carbon literate, and get to grips with the implications not only of our choices but also the bigger infrastructures which underpin the things we consume. How can we educate our desires unless we know what we're choosing between? Mike Berners-Lee, to my complete delight, has provided just the wonderful foundation we need–a book that somehow made me laugh while telling me deeply serious things."

PETER LIPMAN, DIRECTOR OF SUSTRANS

Mike Berners-Lee

THE Carbon Footprint ••• OF ••• Everything

GREYSTONE BOOKS
Vancouver/Berkeley/London

Originally published as *How Bad Are Bananas? The Carbon Footprint of Everything* in North America by Greystone Books in 2011; updated North American edition published in 2022 as *The Carbon Footprint of Everything.*

UK edition originally published by Profile Books in 2010;
updated UK edition first published in 2020 by Profile Books Ltd.

22 23 24 25 26 5 4 3 2 1

Greystone Books Ltd.
greystonebooks.com

Cataloguing data available from Library and Archives Canada
ISBN 978-1-77164-576-8 (pbk.)
ISBN 978-1-77164-577-5 (epub)

Copy editing by Lynne Melcombe
Proofreading by Alison Strobel
Indexing by Bill Johncocks

Cover design and composite by Jessica Sullivan
Cover composite illustration credits: Space Vector, Amanda RY, NIKI90 / Shutterstock

Printed and bound in Canada on FSC® certified paper at Friesens. The FSC® label means that materials used for the product have been responsibly sourced.

Greystone Books gratefully acknowledges the Musqueam, Squamish, and Tsleil-Waututh peoples on whose land our Vancouver head office is located.

Greystone Books thanks the Canada Council for the Arts, the British Columbia Arts Council, the Province of British Columbia through the Book Publishing Tax Credit, and the Government of Canada for supporting our publishing activities.

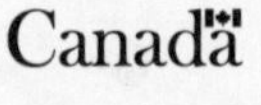

Contents

100 to 1,000 kilos (220 pounds to 1 ton)

1 to 10 tons

10 to 1,000 tons

Millions of tons

Billions of tons

Negative emissions

Introduction

This book has its origins back in 2007. I was working in the nascent field of carbon footprints—working out how much everyday things contribute to carbon emissions—and I agreed to walk around a supermarket with a journalist who was writing about low-carbon food. We trailed up and down the aisles with the Dictaphone running as she plied me with questions, most of which I was pitifully unable to answer. "What about these bananas? How about this cheese? It's organic. That must be better, no? Lettuce must be harmless, right? Should we have come here by bus? How big a deal is food anyway?"

It was not at all clear what a carbon-conscious shopper should buy and on that day we couldn't fill in much of the knowledge. Indeed, the article never happened, and probably just as well. But after that, I began looking long and hard into all kinds of carbon footprints and carried out numerous studies, including one for a supermarket chain. This book, which first came out in 2010 under the title *How Bad Are Bananas? The Carbon Footprint of Everything* and has been completely revised for this new 2020 edition, set out to answer the journalist's questions, and many more besides.

Once I began writing, it was clear that it should be more than just a book about food and travel. I wanted to give a sense of the carbon impact—that is, the climate change impact—of *everything* we consume and do and think about, both at home and at work. I wanted to help us all to develop a *carbon instinct*.

Although I have discussed the footprint of just over a hundred items, I hope that as you read about these you will gain a sense of where carbon impacts come from, so that you will be able to guesstimate the footprint of more or less anything you come across. It won't be exact, but I hope you'll at least be able to get the number of zeros right most of the time. For example, you will have a good idea of how bad bananas are (spoiler alert: not bad at all; they turn out to be a fine low-carbon food, albeit with sustainability issues).

How is this new edition different?

Almost all the numbers have had to be updated for the 2020 edition. But the big difference, ten years on, is that the context has moved a long way forward. In 2010, climate change was just "very serious." Four years had passed since the documentary *An Inconvenient Truth*[1] had catapulted climate change into the American media and popular consciousness, and it was clearly time to start getting carbon awareness into daily life. But today we have a full-on climate emergency on our hands as global emissions have carried on rising as if we'd never noticed a problem.[2]

Meanwhile, the science has been getting notably scary: a temperature change of 1.5°C above pre-industrial levels is now widely acknowledged to be more dangerous than we thought 2.0°C was back then.[3] And we are getting there fast. As I write this, we are at about 1.1°C, compared to 0.88°C in 2010—an increase of 25 percent in ten years. And the effects of climate change are beginning to show around the world: glaciers have shrunk, plant and animal ranges have shifted, trees are flowering sooner, and there's been a dramatic loss of sea ice and accelerated sea level rise. We have had more dramatic climate events, with longer, more intense heatwaves, wildfires, and droughts. Methane has been exploding from the melting permafrost, leaving thousands of craters up to 50 meters (165 feet) across.

That's the bad news. But on the very positive side there is finally a sense that humans might wake up to the challenge. The last couple of years have seen Extinction Rebellion (XR) on the streets and Greta Thunberg leading a global movement of striking schoolkids. We have a long way to go and no time to lose. But compared to 2010 I feel more hope, more fear, and a good deal more urgency.

Revising the entries in the book—and writing new ones on things that weren't on the radar a decade ago, like electric bikes, cryptocurrencies, and the spiraling demands of information technology (IT)—I've tried to keep the tone fun and practical, like the original book. That said, I find it harder to joke about climate these days, and I've become less shy in my messages to policy makers, both in this book and when I'm invited to talk to the media. We need to make it impossible for politicians to pretend they don't understand the essentials of the climate emergency.

At the end of this book, I've added a new section on what each of us can *do* to help deal with climate change. A part of this is about cutting our carbon footprints. The rest has my thoughts on all the other actions we can take to push our governments, workplaces, and society to make the big changes we so urgently need. I'm not trying to tell anyone what to do, but if you are asking, I've got some much more detailed suggestions than I had last time.

Some basic assumptions

The world of carbon counting has moved on slightly in recent years, though it still feels a bit like the Wild West. There's a nasty glitch called "truncation error," which I discuss at the end of this book, that has led government bodies and big companies like Apple, Dell, and HP, to understate their carbon impact by up to 40 percent.

My own basic approach to carbon footprinting, which I practice in an academic capacity at Lancaster University and as a business consultant for Small World, has hardly changed, though I like to think I'm better at it now than I was in 2010. And I hope, at least, that there are three fundamental facts we can agree upon:

- We are in a climate emergency.
- It's human-made.
- We can do something about it.

I hope, too, we can all agree on perspective. A friend once asked me whether his office staff should dry their hands with paper towels or with an electric hand dryer to reduce their carbon footprint. At the same time, he and his colleagues were flying across the Atlantic literally dozens of times a year. A sense of scale is required here. The flying is tens of thousands of times more important than the hand drying, so my friend was simply distracting himself from the real issue.

I want to help you get a feel for roughly how *much* carbon is at stake when you make simple choices, like where you travel, how you get there, whether to buy something, whether to leave the TV on standby, and, of course, where you can get the best return for your effort. This book is here to help you pick your battles. If reading it helps you to think of a few things that can improve your life while cutting a decent chunk out of your carbon, then it'll be a win.

Is carbon like money?

In a sense, yes. Most of the time we know how much things cost without looking at the price tag. We don't have an exact picture, but we know that a bottle of champagne is more expensive than a cup of tea and a lot cheaper than renting an apartment. Our financial sense of proportion allows us to make good choices. If I really want champagne, I know I can have it, provided that somewhere along the line I cut out something just as expensive that is less important to me. Our carbon instinct needs to be similarly attuned.

But that's where the similarity ends. Unlike with money, we are not used to thinking about carbon costs. It's also much harder to tell how much we are spending, because we can't see it and it's not written down. Furthermore, we don't personally experience the consequences of our carbon impact because it's spread across nearly 8 billion people and many years.

Dip in

All of us in the developed world—and I include myself, of course—have plenty of junk in our lives that contributes nothing to the quality of our existence. It's deep in our culture. Cutting that out makes everyone's life better, especially our own. I got a big win by swapping my solo car commutes for bike rides and ride-shares. That works for me, but we are all different.

These pages will, I hope, give everyone some practical and desirable ideas on to how to cut their own carbon footprint and live a better life through carbon awareness.

As to how to use this book, it's designed so you can dip in and flit around. But it's fully indexed and the endnotes often provide further information and links, so I hope it will also work as a reference. Please talk about it with friends and let me know how to improve it next time around (info@howbadarebananas.com).

MIKE BERNERS-LEE
Lancaster, August 2021

A brief guide to carbon footprints

"Carbon footprint" is a phrase that is horribly abused.[1] I want to make my definition clear. Throughout this book, I'm using "footprint" as a metaphor for the total impact that something has. And I'm using the word "carbon" as shorthand for all the different global warming greenhouse gases.

So, I'm using the term "carbon footprint" to mean the *best estimate* we can get of the *full climate change impact* of something. That something could be anything—an activity, an item, a lifestyle, a company, a country, or even the whole world.

What's CO_2e?

Human-made climate change, also known as global warming (or global heating), is caused by the release of certain types of gas into the atmosphere. The dominant greenhouse gas is carbon dioxide (CO_2), which is emitted whenever we burn fossil fuels in homes, vehicles, factories, or power stations.

But other greenhouse gases are also important. Methane (CH_4), for example, which is emitted mainly by agriculture and landfill sites, is twenty-eight times more potent than carbon dioxide, if you compare the impact the two gases will have over a period of one hundred years.

Even more potent, but emitted in smaller quantities, are nitrous oxide (N_2O), which is released mainly from industrial processes and farming and is about 265 times more potent than carbon dioxide over that timescale; and refrigerant gases, which are typically several thousand times more potent than carbon dioxide.

In the US, the total impact on the climate breaks down like this: carbon dioxide (80 percent), methane (10 percent), nitrous oxide (7 percent), and refrigerant and other gases (3 percent).[2]

While these are the factors to apply when you look at the effect these gases have over a one-hundred-year period, the calculations are a bit more complicated, as gases work in different ways. Methane, for example, is much more short-lived than carbon dioxide. This means it does most of its damage in the first ten of those hundred years, by which time the CO_2 has only had about one-tenth of the effect it will have over the course of the century. So, if you are interested in looking at shorter timescales, methane is more than twenty-eight times more powerful than CO_2.

Given that a single item or activity can cause multiple different greenhouse gases to be emitted, each in different quantities, a carbon footprint if written out in full could get pretty confusing.

To avoid this, the convention is to express a carbon footprint in terms of **carbon dioxide equivalent (CO_2e)**. This means the total climate change impact of all the greenhouse gases caused by an item or activity expressed in terms of the amount of carbon dioxide that would have the same impact over a one-hundred-year period.

Beware carbon toe-prints: Direct and indirect emissions

The most common abuse of "carbon footprint" is to miss out some or even most of the emissions caused. For example, many online carbon calculator websites will tell you that your carbon footprint is a certain size based purely on your home energy and personal travel habits, while ignoring all the goods and services you purchase.

Similarly, a magazine publisher might claim to have measured its carbon footprint, but in doing so looked only at its office and cars while ignoring the much greater emissions caused by the printing house that produces the magazines themselves. And countries do this, too, in their carbon calculations, often omitting the footprints of imported goods (from fashion goods to steel and cement) or whole sectors like aviation and shipping.

These kinds of carbon footprint are actually more like carbon "toe-prints," in that they don't give the full picture.

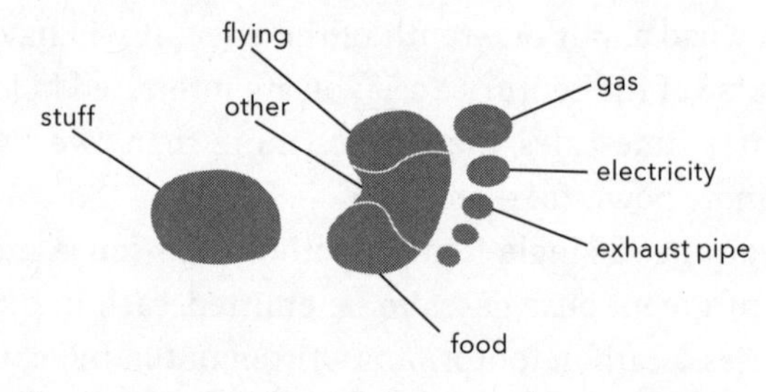

The footprint of a lifestyle is bigger than its toe-print

Much of this confusion comes down to the distinction between "direct" and "indirect" emissions. The true carbon footprint of a plastic toy, for example, includes not only direct emissions from manufacturing and transportation of the toy to the store; it also includes a whole host of indirect emissions, such as those caused by the extraction and processing of the oil used to make the plastic. Tracing back all the things that have to happen to make that toy leads to an infinite (and I am using that word carefully) number of pathways, many of them tiny but important when they are all added together.

To give another example, the true carbon footprint of driving a car includes not only the emissions that come out of the exhaust pipe, but also all the emissions that take place when oil is extracted, shipped, refined into fuel, and transported to the gas station, as well as the substantial emissions caused by producing and maintaining the car.

The essential but impossible measure

The carbon footprint, as I have defined it, is the climate change metric we need to look at. The dilemma is that it is almost impossible to measure. We don't stand a hope of being able to understand how the impact of our bananas compares with the impact of all the other things we might buy unless we have some way of accounting for the farming, transport, and storage processes, as well as the processes that feed into them.

So how should we deal with a situation in which the thing we need to understand is impossibly complex?

One common response is to give up and measure something easier, even if that means losing the real figures you are after. The illusionists Penn and Teller refer to one of their core techniques as the *misdirection of attention*: by focusing the audience on something irrelevant, they can make them miss the bit that matters. This is quite common among companies, or even governments, that declare their carbon footprint. For example, we may find an airport waxing lyrical about the energy efficiency of its buildings while failing to discuss the flights it facilitates. Or a travel company boasting of its sustainable accommodation, again without mentioning flights (flights are often the elephant in the room).

The approach of this book is to make the most realistic estimates that are practical, and to be honest about the uncertainty. I've tried to get the full picture, wherever possible, and above all to get the orders of magnitude clear.

However, huge uncertainty remains and, like so much science, every carbon footprint in this book is a best estimate. So, when you see our number of "6.6 kg (14.5 lbs) CO_2e" for a cheeseburger, it really means "probably between 3 and 10 kg (6.6 and 22 lbs) CO_2e and almost certainly between 1 and 12 kg (2.2 and 26.4 lbs) CO_2e." That is the nature of all carbon footprints. Don't let anyone tell you otherwise.

Some of the numbers are even less certain, especially where I'm trying to bring a sense of scale to important questions that are almost impossible to quantify. Examples include the footprint of having a child, sending an email, or a country going to war. These calculations and assumptions are highly debatable, but I've included them because the thought process can be a useful reflection and because they can still help us to gain overall perspective.

Let me be emphatic that the uncertainty does not negate the exercise. Real footprints are the *essential measure* and nothing short of them will do. The level of accuracy that I have described is good enough to separate out the flying from the hand drying.

And on the subject of flying, one further note is required. For many of us in the developed world, flights represent a significant percentage of our individual carbon footprint. Even if we take a short-haul holiday flight just once a year, that can represent a tenth of our footprint. If we take a long-haul flight, say from New York to London, or Los Angeles to Boston, it may be as much as a half of our footprint. Anyone who takes regular business flights across the Atlantic will likely have a much higher footprint than the average person.

The aviation figures may even be worse than that, as emissions from planes in the sky have a greater impact than those burning the same amount of fuel at ground level. In this book, I have multiplied all aviation emissions by 1.9.[3]

This is possibly a conservative estimate. Some experts believe the true impact of high-altitude emissions could be as much as four times that of regular emissions. (There is a more technical discussion of the methodologies I have used for this on p. 152.)

Making sense of the numbers

So far, we've established what we need to try and measure, but a ton of carbon is still a highly abstract concept.

What does a ton of CO_2e look like?

Well, if you filled a couple of standard-sized garden rain barrels to the brim with gasoline and set fire to them, about a ton of carbon dioxide would be directly released into the atmosphere. (The carbon footprint of burning that gasoline by driving is a bit more than that, for reasons explained later.) If you did the same with a pint milk bottle, that would release just over a kilogram (2.2 lbs) of CO_2, and if you burned a blob the size of a chickpea, that would release about a gram (about a third of an ounce).

1,000 grams (g) = 1 kilogram (kg) or 2.2 pounds (lbs)
1,000 kilograms = 2,200 pounds or 1.1 US tons

To give some sense of scale, the average US person currently has an annual carbon footprint of around 21 tons (down from 28 tons ten years ago). This is similar to the Australians and many people in the oil-producing countries in the Persian Gulf. The average UK person's annual carbon footprint is around 13 tons. For the average Chinese person, it's around 8 tons, and a person in the less developed world has a far lower footprint still. It takes the average US person just a couple of days to clock up the annual footprint of the average Nigerian or Malian. The global average is just over 7 tons per person.

As mentioned earlier, international figures can vary enormously according to the methodology. You get smaller numbers (toe-prints) if you only include the obvious bits of your footprint such as household energy and travel or you miss out emissions like aviation and shipping on goods that are manufactured overseas.

A note about units

When this book was first published in the UK, the measurements were given in grams, kilograms, and tonnes (metric tons). For this North American edition, I have continued to use grams, kilos, and tons of CO_2e as my units for carbon footprints, since that allows us to use a decimal scale that provides a straightforward comparison of impacts. However, I have added some conversions for clarity. In particular, I have given the pound equivalents for measurements in kilograms. So, for example, the carbon footprint of asparagus is described in kg CO_2e per pound.

I have not, however, offered any conversions to most of the measurements in grams. One gram is the mass of about 15 drops of water, and there are 30 grams in an ounce. Because there is such a large disparity between a gram and an ounce, including conversions would lead to unwieldy strings of numbers with a lot of decimal places that detract from the bigger picture.

At the other end of the spectrum are tons, and since a tonne, or metric ton, is reasonably close to a US ton, I've given measurements in tons only, without any conversions. I took the same approach with liters and quarts, which are fairly close, and all pints and gallons mentioned in the text are US pints and gallons.

A 5-ton lifestyle?

To offer a sense of perspective, I have adopted a > 5-ton lifestyle as another unit of measure for this book. In 2010, I used a 10-ton lifestyle, but things have moved on since then and a 5-ton lifestyle now feels more appropriate. It might seem like quite a big challenge for the average US person to cut their carbon footprint down from 21 tons to just 5, but bear in mind that the US is an extremely unequal society. Many people in the US have a carbon footprint much lower than this, but the extremely wealthy drive this average number up by living such carbon-profligate lifestyles. It may still be a significant challenge for most, but in the climate emergency a 5-ton lifestyle feels both possible and necessary. I refer to it from time to time because it gives an alternative and sometimes clearer way of conceiving of those abstract pounds and tons of CO_2e.

There is nothing particularly magic about a 5-ton lifestyle—that is, a lifestyle causing 5 tons of CO_2e per year. It's certainly not a long-term sustainable target for everyone in the world, but if everyone in the US cut to 5 tons right now, it would be a big step forward on the journey to a low-carbon world.

One way of thinking about the footprint of an object or activity is to put it in the context of a year's worth of 5-ton living. For example, a large cheeseburger (6.6 kg or 14.5 lbs CO_2e) represents about an hours' worth of a 5-ton year. If you drive a fairly thirsty gas-powered car for 1,000 miles (1.3 tons CO_2e), that is just over three months' ration. If you leave a couple of the (now old-fashioned) 100-watt incandescent light bulbs on for a year, that would be forty-four days used up. One premium economy return flight from New York to Seoul, South Korea, burns up around 4.7 tons CO_2e. That is nearly a whole year of the 5-ton lifestyle used up in one go, leaving just 300 kg (660 lbs) CO_2e left in the budget for everything else that year: food, heat, buying stuff, healthcare, use of public services, your contribution to the maintenance of roads, any wars around the world that your government is involved in (like it or not)—everything.

You might be wondering whether there are better ways of spending this or any other sized budget than blowing it on burgers, driving, or flying. If that question is of interest, this book is for you.

The world's remaining CO_2 budget

Since, unlike the other greenhouse gases, CO_2 lasts more or less forever in the atmosphere, it is possible to estimate a *total overall budget* for the stuff we can burn in order to stay within any particular temperature limit. This gives us another important comparator to help with that sense of perspective.

Estimates vary, but as of 2018, the *remaining budget* for keeping the world to within 1.5 degrees of warming is about 400 billion tons CO_2. This is a frighteningly small figure, representing a tiny fraction of the emissions we have burned to date and only just over *ten years'* worth of CO_2 emissions at today's levels.

(Remember, too, that CO_2 is not the only calculation in our budget. We need simultaneously to take strong action on all the other greenhouse gases. This is why I've used the wider CO_2e metric for this book.)

Can carbon be offset?

"Offsetting" is a seductive concept, especially when it is offered at prices as low as $4 per ton CO_2e. This would work out to $84 per year for the average US citizen to offset their carbon use and salve their carbon conscience. At that rate, the whole climate crisis could be solved for a trivial 0.2 percent of world GDP (gross domestic product). If only that were true. Sadly, it is nonsense.

All such cheap offset options turn out to be fatally flawed or fundamentally limited in scope. They are often about things like solar power or tree planting that we need anyway to reach carbon zero—we can't just use them to counterbalance our emissions.

The only genuine offset is to remove CO_2 or other greenhouse gases from the atmosphere and store them permanently. These "negative emissions" are expensive and their technologies are mainly in their infancy. They will be needed in our response to climate change, and they are covered in a final section (see p. 209). But there is no substitute for cutting our carbon footprints.

Less than 10 grams

A pint (16 oz) of tap water

0.2 g CO_2e one pint (16 oz) of tap water

33 kg (88 lbs) CO_2e a year's tap water for a typical US citizen

> A year's water for one person is the same as a 53-mile drive in an average US car.[1] That includes drinking, washing, cleaning–all of it

Unlike the bottled alternative, which has over 1,000 times the impact (see *A liter (32 oz) bottle of water*, p. 59), cold tap water is not a major carbon concern for most people. In the US, the provision of household water accounts for about 0.2 percent of the country's carbon footprint.[2] However, if a pint of tap water is poured down the drain, its footprint triples to 0.5 g because it is more carbon intensive to treat waste water than to supply the water in the first place.[3] If the eventual fate of the drink is to be flushed down the toilet along with another 6 quarts, that takes the total to about 7 g CO_2e.

While tap water doesn't have a huge footprint, climate change is now causing serious water stress in many places. After three years of drought, Cape Town only avoided running out of water in 2018 by restricting water use to just 50 liters (13 gallons) per person per day (the average US citizen uses around 82 gallons per day). In the US as a whole, it looks unlikely that there will be shortages of water, even though some redistribution might be called for.

Tap water is one thing, but heating it up is another matter, accounting for a decent chunk of the typical person's emissions (see *A shower*, p. 50, and *Desalinating 1,000 liters (264 gallons) of water*, p. 97).

An email

0.05 g CO_2e spam email picked up by your filters
0.2 g CO_2e short email going from phone to phone
0.4 g CO_2e short email sent from laptop to laptop
17 g CO_2e long email that takes 10 minutes to write and 3 minutes to read, sent from laptop to laptop
29 g CO_2e an email that takes you 10 minutes to write, sent to 100 people, 99 of whom take 3 seconds to realize they should ignore it and one of whom reads it.[4]

> Our average email traffic is equivalent to driving 10–128 miles in a small gas-powered car

The footprint of an email comes from the electricity needed to power the kit used at each stage of the process: the device it is written on, the network that sends it, the data center it is stored on, and the device it is read on. The devices at each end are the dominant factors, even if you send big attachments. As the pie chart overleaf shows, the embodied emissions of a smartphone represent 76 percent of a short email's carbon footprint. That percentage would be higher for a laptop, and a step up again for a desktop computer. (For more on the footprint of buying and using a smartphone, see p. 126, and for a computer, p. 140.)

In 2019, the world's 3.9 billion email users sent 294 billion emails each day, of which 55 percent were spam.[5] So the average email user received about 75 emails per day (of which 41 were spam). On the one hand, if you received this number, with all the non-spam being emails that took the sender just 10 seconds to write and you a mere 5 seconds to read, then the carbon footprint of writing, sending, and reading would be around 3 kg (6.6 lbs) CO_2e per year, or 12 million tons CO_2e globally. On the other hand, if they were all more thoughtful emails, taking the sender 3 minutes to write and you a full minute to read, it would come to 38 kg (84.5 lbs) per year, or 150 million tons globally.[6] In this case, email would account for about 0.3 percent of the world carbon footprint. Luckily, this is not the case.

Although most incoming emails sent are spam, these messages account for only around 2 percent of the total footprint of your email account because, although they are a pain, you deal with them quickly. In fact, you never even see most of them if you have decent filters installed. A genuine email has a bigger carbon footprint simply because it takes more time to deal with. So, if you are someone who needlessly copies people in on messages just to cover your own back, the carbon footprint is one more good reason to change your ways. You may find that after a while everyone at work starts to like you more, too.

The long email sent from a laptop has one-twentieth the footprint of a letter (see p. 17). That looks like a carbon saving unless you end up sending 30 times more emails than the number of letters you would have mailed in days gone by. Lots of people do, and perhaps still send the occasional letter as well. This is a good example of the *rebound effect*—how a more efficient technology typically results in higher-carbon living because our usage goes up by even more than the efficiency improvement.

If the great quest is for ways in which we can improve our lives while effortlessly cutting carbon, surely spam and

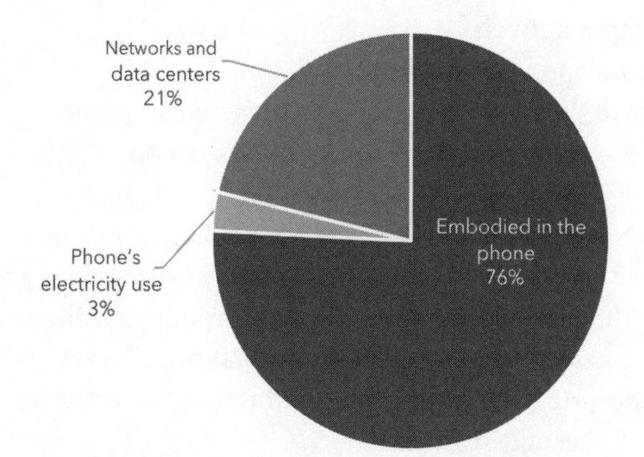

Total carbon footprint of a short email sent from phone to phone over Wi-Fi and taking 10 seconds to write and 5 seconds to read

unnecessary email have to be very high on the hit list, along with old-fashioned paper junk mail. In 2019, OVO Energy ran a campaign to stop people sending needless "thank you" emails. I supported it as a great way into a bigger conversation about our climate emergency, with the realization that there is carbon in everything and there are benefits to cutting every kind of junk out of our lives. But, of course, the actual carbon to be saved from reducing the smallest emails of all is tiny, and it can be wonderfully important to say thank you!

If only email were taxed. Just a penny per message would surely kill most spam. The funds could go to tackling world poverty, or even renewable energy. The world's carbon footprint would go down by 2.4 million tons,[7] the average user would save a couple of minutes every day, and there would be a $480 billion annual fund. If one cent turned out to be enough to push us into a more disciplined email culture with perhaps half the emails sent, the anti-poverty fund would be cut in half, but our lives would still be significantly better. The (small) carbon saving would be an additional bonus.

A Google search

0.6 g CO_2e one simple search
6.9 g CO_2e 5 minutes web browsing from a smartphone
9.6 g CO_2e 5 minutes web browsing from a laptop[8]

> It's good to stay informed

Based on Google's estimate for the energy used at their end (and adding a bit for your phone or computer and the network), a simple web search is about 4 seconds' worth of a 5-ton footprint, while a 5-minute browse on your laptop is around 60 seconds' worth.

To get these figures, I started off with Google's estimate from 2009 of 0.2 g CO_2e for the electricity they use at their end when you put in a single search and guesstimated this is now twice as efficient.[9] I added 30 seconds' use of a smartphone while you tap in the search, wait for the result, and scan it for what you want, including the energy used and the embodied carbon; the network, assuming you use mobile data, adds another 0.5 g, bringing the total to 0.6 g. For the high-end figure, I assumed use of a reasonably efficient laptop, which uses more power than the phone and, more importantly, has a much higher embodied footprint in its manufacture. Almost a quarter of the laptop search comes from the Wi-Fi.

If you search for information about the footprint of web searches, you'll find blogs and articles all coming up with different figures based on different assumptions. Researchers don't always agree, but the figures here should be in the right ballpark.

Google is estimated to deal with 3.5 billion searches per day—up from 200–500 million in 2010.[10] If we go with the figure for the footprint of a single search on your smartphone over mobile data, Google-searching accounts for almost 630,000 tons CO_2e per year. That sounds like a big number but it's less than 0.0001 percent of humanity's carbon footprint. We can probably relax about it. Reading the stuff we find, however, is an altogether more carbon-hungry activity (see A *computer (and using it)*, p. 140).

A text message

0.8 g CO_2e single text message[11]

> A text is no big deal–the world's 9.5 trillion texts account for just 0.01 of global emissions

Around the world, about 9.5 trillion texts are sent every year.[12] The most obvious part of a text message's carbon footprint is the power used by your phone while you type and by your contact's phone while they read what you've written. If you take a minute between you to type and then read the message, and you each have phones that consume 2 watts of power when in use, the footprint of the electricity used will be about one-fiftieth of a gram. However, this is only about 2 percent of the story. Again, the main part of the footprint is the embodied carbon in the phones themselves, and the fact that you wear them out a little bit every time you use them (see p. 126). The transmission of a 140-character message across the network turns out to be tiny—about 0.001 grams.

The average American adult sends 15 text messages per day, adding up to roughly 1.8 trillion texts per year (a lot more than the average UK mobile user, who only sends only 2.5 texts per day).[13] People are also increasingly shifting to online messaging apps. For example, WhatsApp, Facebook Messenger, and WeChat all have very similar footprints to text messages, although the network footprint is a tiny bit bigger. But, overall, they also come to 0.8 g per message if each one takes you 30 seconds to write and the receiver 30 seconds to read.

And what does this all add up to? A 7-million-ton global footprint, which sounds a lot but is just over 0.01 percent of the world's carbon footprint. In other words, texting is not a big deal.

A plastic grocery bag

3 g CO_2e very lightweight variety
10 g CO_2e heavier supermarket bag[14]
50 g CO_2e heavyweight reusable bag

> Plastic bags are bad for many reasons, but carbon isn't chief among them

Plastic is a dazzling example of how we humans invent things and start using them without really understanding the impacts. But, at last, we seem to have woken up to the hideous plastic pollution problem that we have unthinkingly inflicted on the world over the last fifty years. Since the first edition of *Bananas*, the UK government (and many others) has applied a mandatory plastic bag levy, and this has sent such a signal that everyone is cutting down on plastic bags, even people who are not worried about either the cost or the environment. In the US, there is no plastic bag levy on a national scale, although eight states (California, Connecticut, Delaware, Hawaii, Maine, New York, Oregon, and Vermont) have introduced bans on disposable plastic bags, as have several US cities.[15]

All this is good, but has it helped us come to grips with our climate emergency? Not really.

Plastic bags have a pretty low carbon footprint. If you use six old-fashioned bags per week, it works out to about 3 kg (6.6 lbs) per year—the carbon footprint of eating just half a hamburger. Or, to put it another way, when you carry your shopping home in a disposable plastic bag, the bag is typically responsible for about one-thousandth of the footprint of the food it contains. (Note that if you get fewer than five uses out of a heavy reusable bag, you'd be better off, in carbon terms, with disposables.)

Of course, there are other good reasons to ditch single-use plastic bags. Plastic has a habit of hanging around in the ecosystem, where it can sit for hundreds of years, killing fish and being ugly. When we talk about it degrading over time, all we really mean is that it falls into smaller and smaller pieces; as far we know, we are stuck with it forever. And we use an awful lot of it. If all the world's discarded plastic were cling wrap, you could wrap the world up one-and-a-half times.[16]

How to get rid of plastic, then? Burning releases nasty toxins, as well as carbon, although the technology is improving. From a purely climate change perspective, landfill is not too bad. The bags won't degrade, so all those hydrocarbons are returned to the ground where they came from for fairly long-term storage. But landfill is nasty for other reasons (see 1 kg (2.2 lbs) *of trash to landfill*, p. 71).

Drying your hands

0 CO_2e letting them drip
3 g CO_2e Dyson Airblade
10 g CO_2e one paper towel
22 g CO_2e standard electric dryer

> If you use office washrooms six times per day, your hand drying could produce 6–47 kg (13–103 lbs) CO_2e per year

"What's the greenest way to dry my hands?" is a frequently asked question, so I'll answer it, even though (as I mentioned in the introduction) if you want a lower-carbon lifestyle, you really should be asking about something more important.

At the low-carbon end of the scale is drying your hands with a Dyson Airblade. This dryer does the job in about 15 seconds with 1.6 kW of power.[17] Its secret is that it doesn't heat the air, it just blows hard. This makes it far more efficient than conventional hand dryers.

At the high end are paper towels and conventional heated hand dryers. The paper towels are based on 10 g of low-quality recycled paper per sheet and only one towel used each time.[18] (Of course, if you use two towels the footprint doubles.) Conventional hand dryers are around four times worse than the Dyson because they take a shade longer and use around 6 kW of power (it takes a lot of energy to create heat).

Right at the bottom of the scale comes not drying your hands at all, or using a small hand towel that is reused many times between low-temperature washes. I am not a hygiene expert, but I'm told that neither option is good from that point of view, and certainly not in multi-user washrooms during a pandemic. They may even end up adding to the already substantial footprint of healthcare (see p. 146).

Finally, drying hands usually follows washing them. For this, warm water is higher carbon but useful for hygiene reasons, and typical US mixer taps are vastly superior to the traditional British twin taps that involve you filling the whole sink or dancing your hands between the two and trying not to get scalded.

10 to 100 grams

A paper bag

12 g CO_2e recycled and lightweight

80 g CO_2e a clothing store bag from mainly virgin paper

> It's always better to reuse bags

A common misconception is that paper bags must be both more ecological and lower carbon than plastic. The carbon assumption is wrong. The paper industry is highly energy intensive. Printed virgin paper typically produces 2.5–3 kg (5.5–6.6 lbs) CO_2e per kilo of paper manufactured (or 1.1–1.4 kg per pound). This is comparable to the emissions required to produce 1 kg (2.2 lbs) of polypropylene plastic bags. However, paper bags have to be much heavier, so overall the paper bag ends up having a bigger footprint.

Recycled paper is roughly half as energy intensive to produce as virgin paper. But even a lightweight recycled paper bag, which doesn't always work with fruit and vegetables, produces slightly more greenhouse gas emissions than a typical plastic bag.

There is another problem at the disposal end, which I have not factored into my numbers. Unless you recycle, which of course we all should, your paper bag is likely to end up in a landfill, where it will rot, emitting more CO_2 and, even worse, methane. Landfill sites vary in their ability to capture and burn methane emissions, but none of them capture it all and typically there will be around 1 kg (2.2 lbs) of greenhouse gas emissions per kilo of paper buried (see *1 kg (2.2 lbs) of trash to landfill*, p. 71).[1]

Just as with plastic bags, it's always best to reuse paper bags, and always good to recycle them. It's better still to avoid them by using your own reusable bags, which have the added advantage that in wet weather they won't fall apart and set your bruised apples rolling down the street.

Ironing a shirt

15 g CO_2e quick, expert skim on a slightly damp shirt
27 g CO_2e average
76 g CO_2e thoroughly crumpled shirt

> Ironing five shirts a week for a year is about the same as a 9-mile drive in an average car

A friend of mine used to iron her husband's socks (they're now divorced). If you're feeling stuck in a similar routine, I hope you will find the carbon argument gives a bit more power to your elbow. Although ironing isn't the biggest environmental issue, there may be scope for saving a little bit of carbon here, and perhaps some lifestyle improvement, too. For ironing that simply has to be done, the greenest step is to have the clothes slightly damp and use the ironing process itself to finish off the drying. That saves both time and carbon, especially if you would otherwise be using an energy-hungry tumble dryer (see p. 96).

Even more effective is simply using the iron less often. A few people allegedly iron almost as a hobby or a form of meditation. If that's you, then I have some good news: improvements in the UK electricity mix mean that ironing now works out at about 140 g CO_2e per hour, well down from the 250 g CO_2e per hour when the first edition of this book came out. It's actually now comparable to watching an hour of TV, which turns out to be worse than we thought it was back in 2010, though not so good, of course, if you do both at the same time.

A Zoom call

4 g CO_2e per hour on a 13-inch MacBook Pro
20 g CO_2e per hour on a laptop of average efficiency
97 g CO_2e per hour on a desktop computer with screen
+ embodied emissions in the computer
(-) 22 tons CO_2e potential saving on a meeting in Hong Kong with one person flying from London and one flying from New York

> A video call can save several commutes or tons of carbon from several flights

The chief footprint of a Zoom (or Teams/Skype/FaceTime/Google Hangouts) call is the embodied emissions from your hardware. The actual call is pretty minimal and not much different than regular computer use (see p. 140). But the emissions you *avoid* can be of a whole different order.

Let's say there are four of you traveling an average 5 miles to a meeting. If you're all driving, that could net 16 kg (35 lbs) of emissions, more if one of you has a gas-guzzling SUV. But what if it's a meeting in Hong Kong and you're flying there from New York and someone else is coming from London? If both of you travel return by business class, that would accumulate 22 tons CO_2e. Now imagine that Zoom call replacing a conference where a company or several universities sent 200 or more delegates.

The downside of video conferencing comes if you weren't thinking of flying to meet someone at all, but after a few calls you decide you are getting on so well that you simply have to meet up. Here, yet again, is the ugly face of the rebound effect. It is evidenced in this case by the uncomfortable reality that, up until the COVID-19 crisis, the rise of video conferencing had simply gone hand in hand with ever-rising air traffic for both business and leisure.

As we all know, video conferencing mushroomed during the COVID-19 crisis, as offices closed and flights were grounded. The carbon footprints of commuters and businesses improved. Many businesses found they were able to function pretty well holding many of their regular meetings online. Although pure Zoom-life can get tiresome, the world has learned that it is both possible and better to replace many needless, high-carbon, time-consuming trips with a simple video call. In this respect, we will surely never look back.

A 100 g (3 oz) serving of carrots

28 g CO_2e local, in season, full-size varieties

83 g CO_2e local, in season, baby carrots

95 g CO_2e full-size varieties, traveled by road from California to Texas

> Carrots are about as good as it gets from a carbon and nutritional point of view

The figures above are for 100 g of carrots, a decent-sized serving, and the figures are good for most root vegetables. At around 0.7 g CO_2e per calorie, local root vegetables are some of the most climate-friendly foods available, and healthy too. If you ate only these foods, and others with a similar carbon intensity, you could feed yourself for about 550 kg (1,210 lbs) CO_2e per year.

Seasonal vegetables have small carbon footprints because they are grown in natural conditions without artificial heat, they don't go on planes, and they don't incur inefficiencies similar to the production of food from animals. Note that some baby varieties have a much lower yield per acre of land, resulting in higher emissions per kilogram. And, as with other vegetables, favoring misshapen specimens may help avoid waste in the supply chain (see *An apple*, below).

If you boil your carrots for 10 minutes, you will add a few more grams CO_2e per kilo to the footprint (for more on cooking, see *A 200 g (7 oz) serving of boiled potatoes*, p. 38). My children only used to eat their carrots raw, which suited me fine. It was better from every angle—fewer carbon emissions, less time, and a higher nutritional value.

An apple

0 CO_2e plucked from the garden
32 g CO_2e local and seasonal
110 g CO_2e long distance road, seasonal
320 g CO_2e long distance road, out of season, frozen

> Apples are low-carbon food wherever they come from but local and seasonal is best[2]

The statement above seems obvious and incontrovertible, but to give you an idea of the complexity and uncertainty of calculating carbon footprints, consider this. A New Zealand university study claimed to have found that their country's apples exported to the UK market had a footprint of just 185 g CO_2e per kilo (or 84 g CO_2e per pound)—significantly lower than UK apples for local consumption, which came in at 271 g per kilo (or 123 g CO_2e per pound).[3] The argument was that UK production entailed greater use of fossil fuels on the farm and required more cold storage. The study also pointed to the cleaner electricity mix that New Zealand had at the time compared to the UK. These factors, it claimed, outweighed the emissions from shipping the produce halfway around the world. A similar study by the UK government's Department for Environment, Food and Rural Affairs (Defra) produced similar orders of magnitude but found, conversely, that for Germany local apples were more carbon friendly than those from New Zealand.[4] It's difficult to determine the truth. Each study made slightly different assumptions.

There is no question that local, in-season apples are best. However, in early summer, when any local apples will have been in cold storage for months, importing may be the lower-carbon option. But there is nothing particularly bad about buying them from anywhere in the world because they travel on boats rather than by air. And one other point: as with all fruit and vegetables, it's a good idea to buy the most misshapen ones you can get, because that encourages the supply chain not to chuck them in the trash before they reach the store.

A mile by electric bike

5 g CO_2e per mile for a fully electric bike traveling at 12 mph with no hills or stops

9 g CO_2e per mile at the same speed with five stops per mile and 20 meters of climbing[5]

+ 10-100 g CO_2e per mile for the bike's embodied carbon

> Electric bike emissions are truly amazing

In the ten years since the first edition of this book, electric bikes have come of age—and they turn out to be incredible from a carbon point of view. They are so good that I am just about to buy a folding one for my daily commute.

How can a mile on an electric bike be twenty times more carbon friendly than a mile on a conventional machine? The astonishing figures begin to make sense when you consider how you get your own energy—that banana trees are many times less efficient at capturing the sun's energy than, for example, solar panels. Also, only a small part of the banana tree's energy finds its way into the banana itself, which then has to be transported around the world to reach you. On top of this, electric motors are perhaps four times as efficient as human legs at turning chemical energy into bike propulsion. If all our electricity was from solar power, the electric bike would beat the conventional bike by a factor of nearly 1,000.

My numbers are for a fully motorized ride, although, by law, electric bikes must be a hybrid between pedaling and motor power. They assist you, but don't do all the work. This is probably just as well, because it means that electric bikes still keep you fit but make possible a greater range. However, this also means that the real footprint of cycling an e-bike is somewhere between the footprint of conventional cycling and the numbers I've given here.

The embodied emissions in the e-bike are similar to a regular bike except for the extra battery and engine. But they are probably lower per mile, since you are likely to ride the bike farther over its lifetime than a conventional bike. The battery turns out to account for just 0.5 g per mile provided you use it to the end of its life.[6] I've assumed that you will look after your battery carefully, so that you get 1,000 full charges out of it. To do this, charge it up slowly (trickle charge) and neither let it run totally flat nor charge it to the very top. And don't let your bike stand around for weeks on end—get out and use it![7]

Walking through a door

0 CO_2e a normal household door on a summer's day
3 g CO_2e your front door on a cold winter's day
83 g CO_2e big electric doors opening into a large stairwell on a cold, windy day[8]

> At the high end, that's a banana's worth of carbon every time you enter a building

The entrance door of the building where I work has no manual option. To get in, you have to press a button and wait while two electric motors whir and double doors swing slowly open, creating a space 2 meters wide by 2.5 meters high. You enter a spacious stairwell with two large radiators. The only decoration is a certificate proclaiming the D-rated energy performance of the building. It takes 18 seconds for the doors to finish closing. This twelve-year-old building was amazingly rated environmentally Excellent in its BREEAM evaluation.[9]

The power used by the electric motors isn't the problem. They generate less than half a gram of CO_2e. The problem is the size of the space you have to open (since the first edition of this book came out, one side of my building's doors stays shut, slightly mitigating the problem), the time it has to stay open for, and the vast heated space that the doors open onto. For this building, there must have been lots of other options, such as manual doors that swing shut and can be opened singly, with an override button for disabled access. Rotating doors attached to turbines that generate electricity as you pass through have been trialed in Holland[10] but sound like the kind of gimmick that can tarnish the reputation of the renewables industry.

In a typical home on a cold, blustery day, the numbers are more likely to come out at about 3 g, based on opening a door by hand and closing it immediately.

However inefficient the process, and however poor the design, on a cold day it is still far better to open a door and close it than to keep it open all the time. Shops that keep their doors open in winter are doing so regardless of the carbon consequences.

Boiling a liter (a quart) of water

50 g CO_2e using a gas kettle on fairly low heat
75 g CO_2e using an electric kettle
115 g CO_2e using a saucepan on a gas stove without a lid

> Gas kettles are still best for now, but electric will be better with lower carbon electricity generation

Some friends of ours have a stovetop kettle that they use on their gas stove, and we ended up debating the environmental pros and cons for months. Finally, I spent half a morning measuring different methods. (A sad way of spending time, I know, but I did have a book to write.) It turned out that our plug-in electric kettle was the fastest, and only 10 percent of its energy was wasted. So, although inefficiencies in our power stations and distribution systems make electricity a high-carbon way of producing heat, the electric kettle is still a fairly good way of boiling water at home. The way the electricity is generated also makes a difference. If the electrical grid in your country relies more on renewable energy and less on fossil fuel, the electric kettle will always win (as is the case in the UK and some other countries already).

How the gas kettle compares with the electric kettle depends on the time of year. In winter, our gas-using friends easily win the low-carbon prize. That's because, although some of the heat from the gas flames escapes around the edge of their kettle, that heat wasn't actually wasted: the kitchen is the heart of their house, so all the heat going into the room was useful. In their house, in fact, the gas stove was the most efficient form of heating because nothing was wasted up the flue (as in a gas furnace), sent to unoccupied rooms, or lost in pipework (as in central heating).

In the summer, our friends still won the low-carbon prize provided they were willing to put their kettle on a small gas ring to maximize the proportion of the heat that went into the water, rather than being lost around the sides. Doing this gave them a 30 percent carbon saving over the electric kettle but also meant it took three times as long (twelve minutes) to boil. If they used a large gas ring, the result was slightly *more* carbon than the electric kettle—and it was still 50 percent slower.

Saucepans turned out to be less efficient than kettles. It only makes sense to bring water to the boil in a saucepan if you are putting vegetables in at the start, in which case there is the benefit that they begin cooking a bit even before the water boils. If you do use a saucepan, keep the lid on (20 percent waste if you don't) and make sure the flames don't go up the sides (potential for another 20 percent waste). Just as important is not to boil more water than you actually need.

A couple of sustainable kettle design features are worth a mention, since there are some incredibly simple features that have taken a long time to become widely available. First, good insulation cuts the heating time, saves carbon, and means that if you accidentally boil more than you need it stays hot for longer. Second, a thermostat allows you to set a kettle to 85°C (185°F), which is all you want for herbal and arguably regular tea. This is quicker, cheaper, lower carbon, tastier, and probably reduces the chance of mouth cancer.

Traveling a mile by bus

11 g CO_2e on a full 90-seater electric bus in the US
26 g CO_2e squeezed into a minibus in La Paz, Bolivia
198 g CO_2e on a half-full single-decker bus
6 kg (13.3 lbs) CO_2e per mile single-decker bus shared with just the driver

> The efficiency of any given bus is proportional to the number of people it is carrying

It also depends on the amount of stopping and starting and, of course, the energy source. A standard US city bus will have a footprint of 99 g CO_2e per passenger mile when full. If the bus is only half full, the footprint per passenger mile doubles. If it is just you and the driver, it jumps up to 6 kg (13.3 lbs) CO_2e per passenger mile. But another way to look at it is that the bus was going anyway, and you getting on the bus is just about carbon-free. It's a Catch-22. No one wants to take the bus, because it is just as cheap and quicker to take the car (if you've got one). However, if the bus were three times as frequent and one-third of the price, it would probably be very popular.

La Paz, Bolivia, is the place I think of where this principle is practiced to perfection, provided you are prepared to set aside a bit of safety and comfort. Twelve-seater minibuses charge around town with twenty or more people crammed inside. You can get just about anywhere for one boliviano (a few pennies) and you are unlucky if you have to wait more than five minutes. Most people in the developed world would choose a luxury version of this for perhaps five times the price, but the principle is sound and in Bolivia twenty years ago the value proposition met the market need perfectly.

All my numbers have factored in the fuel supply chains as well as the exhaust pipe emissions. I have also included a component for the emissions entailed in manufacturing the vehicle, although for the bus this is a small consideration because they do so many miles before needing replacement.[11]

Going forward, of course, all buses need to be electric. This will also help cut back on the 90,000–360,000 premature deaths per year in the US from air pollution.[12]

Cycling a mile

40 g CO_2e powered by bananas
70 g CO_2e powered by cereal with cow's milk
190 g CO_2e powered by bacon
640 g (1.4 lbs) CO_2e powered by cheeseburgers
3.2 kg (7 lbs) CO_2e powered by airfreighted asparagus
+ 10–100 g CO_2e per mile for the bike's embodied carbon[13]

> If your cycling calories come from burgers, the emissions are about the same as driving

I have based all the above calculations on the assumption that you burn 50 calories per mile. The exact figure depends on how fit you are (the fitter, the lower the emissions), how tall, wide, and heavy you are (the more, the higher the emissions, as they add to the air and rolling resistance), how fast you go (the faster, the higher) and how much you have to use the brakes.

All that energy on a bike has to come from the food you eat, which in turn has a carbon footprint. The good news is that the lower-carbon options are also the ones that make the best cycling fuel. Bananas, of course, are brilliant (see p. 40). Breakfast cereal is pretty good (let down slightly by cow's milk). The bacon comes in at around 190 g CO_2e for a

25 g slice, with only enough calories for a mile and a quarter of riding, while cycling along using calories from cheeseburgers is equivalent to driving the same distance in an average car. At the ridiculous end of the scale would be getting your cycling energy by piling up your plate with asparagus that has been flown by air from the other side of the world. At 3.2 kg CO_2e per mile, this would be like driving a car that does 4 miles to the gallon (a shade over a mile per liter). You'd be better off in an SUV.

The emissions embodied in the bike and requisite equipment vary per mile depending on how much use you get out of it and whether you buy new or secondhand. In the lower-carbon scenarios, the food accounts for only a small part of your impact, and the maintenance of your bike and sundry equipment dominates.

Is cycling a carbon-friendly thing to do? Emphatically, yes! Powered by low-carbon carbs, a well-used and well-maintained bike is about ten times more carbon efficient than the average gasoline car. Cycling also keeps you healthy, provided you don't end up under a bus. (Strictly speaking, dying could be classed as a carbon-friendly thing to do, but needing an operation couldn't: see p. 146.)

Buying a folding bike so I could commute on the train has been one of the best decisions I ever made, in terms of both lifestyle and carbon. My journey takes ten minutes longer, but I get half an hour's exercise and fifteen minutes reading a book each way. So, I've magicked an extra hour of the stuff I love into my day while saving money and carbon. And, by taking my car off the road in rush hour, I cut everyone else's queuing time as well and reduce the emissions they belch out while they wait (see A *rush-hour car commute*, p. 114).

100 to 500 grams (3.5 to 17.5 oz)

A 200 g (7 oz) serving of boiled potatoes

56 g CO_2e locally grown, raw
106 g CO_2e locally grown, boiled gently, lid on
240 g CO_2e traveled by road from Idaho to New York, boiled gently, lid on
350 g CO_2e traveled by road from Idaho to New York, boiled furiously, lid off

> With potatoes, local is best but it's mainly about the cooking

If local boiled potatoes were all you ate for a year, you could feed yourself for just 280 kg (616 lbs) of CO_2e, or just under 6 percent of the 5-ton lifestyle. That is pretty good when you consider that food and cooking currently accounts for around 4 tons of CO_2e per person per year in the US. Of course, you'd end up bored and malnourished if you stuck to this regimen, but there is clearly a place for potatoes in the low-carbon lifestyle. I've calculated the figures on a fairly generous 200 g portion, which would represent about 8 percent of the average person's daily caloric needs.

Potatoes are a low-carbon crop; larger conventional varieties are especially good, as yields are higher. Transport emissions are not high, either, provided potatoes stay local. However, it is not uncommon for supermarkets to move produce hundreds of miles to a distribution center and then back again. But, even when this happens, the transport does not have a disastrous impact.

The biggest part of the footprint comes from the cooking process. Do this efficiently and you can cut your carbon impact in half. Here are some ways to keep the emissions to a minimum:

- **Use a lid on the pot**
- **Boil gently** The temperature of the water, and therefore the cooking speed, is exactly the same when you turn the heat down to a gentle simmer as when you cook it at a full boil.
- **Cut the potatoes into smaller pieces**
- **Use a pressure cooker** The pressure raises the boiling temperature, which means the potatoes cook faster and more efficiently.
- **Ideally, use an induction stove**

Alternatively, if you are baking or roasting, you can do the following:

- **Use a microwave or a fan-assisted gas oven**
- **Having heated the oven up, cook more than one thing**

I have ignored in this entry the carbon cost of getting to the store (see *Driving a mile*, p. 67, and *Cycling a mile*, p. 36).

Local potato footprint (200 g serving)	**Grams CO_2e**
Growing the potatoes	44
Transport	10
Packaging in a simple bag	2
Supermarket storage and display	2
Boiling	50–160
TOTAL	106–216

A banana

110 g CO_2e each (or 300 g CO_2e per pound)[1]

> To answer the question in the original title of this book: bananas aren't bad at all

Bananas are a great food for anyone who cares about their carbon footprint. For just 110 g of carbon, you get a whole lot of nutrition: 143 calories as well as loads of vitamins C and B_6, potassium, and dietary fiber. Overall, they are a fantastic component of the low-carbon diet. Bananas are good for just about everyone—athletes, people with high blood pressure, everyday cycle commuters in search of an energy top-up, or anyone wishing to chalk up their recommended five portions of fruit and vegetables per day. There are three main reasons that bananas have such low-carbon footprints compared with the nourishment they provide:

- They are grown in natural sunlight with no hothousing.
- They keep well, so although they are often grown thousands of miles from the end consumer, they are transported by boat (which is about 1 percent the carbon footprint of airfreight).
- There is hardly any packaging because they provide their own. Supermarkets sometimes feel they have to put bananas in plastic bags to stop customers ruining them if they split a bunch, so the bag can be worth it in preventing waste.

On top of their good carbon and healthy eating credentials, the fair-trade version is readily available. However, for all their qualities, don't let me leave you with the impression that bananas are too good to be true. They have environmental issues. Of the 300 types in existence, almost all those we eat are of the single, cloned Cavendish variety. The adoption of this monocrop in pursuit of maximum, cheap yields has been criticized for degrading the land and requiring the liberal use of pesticide and fungicide. And it has left bananas vulnerable to the *Fusarium* fungus, which has been slowly spreading around the world, devastating plantations.

Furthermore, although land is dramatically better used for bananas than beef in terms of nutrition per acre, there are still parts of the world in which forests are being cleared for banana plantations (see *Deforestation*, p. 191).[2]

Overall, though, the only really bad bananas are those you let rot in your fruit bowl. These join the scandalous 32 percent of food wasted by consumers in the US and many other countries.[3] If you do find yourself with bananas on the turn, they are good in cakes and smoothies. I have a childhood memory that they are also tasty in pudding.

A diaper

93 g CO_2e reusable, line-dried, washed at 60°C (140°F) in a large load, passed on to a second child

130 g CO_2e disposable[4]

300 g CO_2e reusable, tumble-dried and washed at 90°C (195°F)

> That's 490 kg (1,078 lbs) per child in disposables—the equivalent of nearly 900 lattes

On average, reusables come out at 600 kg (1,320 lbs) CO_2e per child compared with 490 kg (1,078 lbs) for disposables. At their best, reusables can have a footprint which is less than that of disposables. But to achieve this you need to pass them on from one child to another (so that the emissions embodied in the cotton are spread out more), wash them at a lower temperature (60°C/140°F), hang them out to dry on the line, and wash them in large loads. If you wash them very hot and tumble-dry them, reusables are worse than disposables.

For a disposable diaper, most of the footprint comes from its production. But up to 20 percent arises from the methane emitted as its contents rot down in landfill (it's a myth that if you wrap them up in a plastic bag they will never rot). Biodegradable disposables cause less plastic pollution but are worse for methane if they end up in landfill.

The study I'm basing my figures on for reusables assumed that the average child stays in diapers for about two-and-a-half years, and is changed just over four times a day, so uses about 4,000 diapers in total.[5]

What does all this mean for the eco-conscious family? If you have two children and stick to line-dried reusables throughout, you might be able to save more than 275 kg (605 lbs) of CO_2e. You will also cut out landfill and the hideous problem of discarded plastic pervading the world. It's a significant benefit, but (here's the catch) you need to know your own minds before you start out, because if you give up, revert to disposables, and discard the reusables, it could be the option with the highest footprint of all.

To keep all of this in perspective: if you take just one family holiday by plane, you will undo the carbon savings of perfect diaper practice many times over.

An orange

150 g CO_2e each (or 440 g CO_2e per pound)

> Oranges are great but juice is not so good

Most oranges, along with most apples and bananas, are great from a carbon perspective.[6] They keep well, so they can be grown in natural conditions and shipped around the world. The important thing to note is that although there are lots of food miles, they are fairly climate friendly. Like bananas, oranges can go on a huge boat and take their time. However, some supermarkets airfreight some varieties of orange at the start of the season to get them into the stores a couple of weeks early.

Orange juice is not nearly as good as just eating the fruit. I estimate that a quart of orange juice has a footprint equivalent to just over 6 kg (13.2 lbs) of oranges—many more than it would take to produce that much juice—as juice incurs several inefficiencies in its production:

- The pulp is thrown out (so varieties with pulp and smoothies made from fresh-squeezed oranges may be more sustainable).
- There are emissions from processing, including pasteurizing and turning it into concentrate for transport and refrigeration.
- There is the footprint of the carton.
- Transport miles are often higher, as the product moves from farm to juicer to cartoner to distributor around the world.
- Fresh orange juice requires refrigeration. Tesco (a British supermarket chain similar to Safeway or Albertsons in the US) reported once that their freshly squeezed juice has about twice the footprint of the long-life product.

A supermarket delivery

40 g CO_2e for a supermarket shop by electric bike
150 g CO_2e for a supermarket delivery by electric van
450 g (1 lb) CO_2e for a supermarket delivery by diesel van
1.2 kg (2.6 lbs) CO_2e if you drive yourself in an average gas-powered car

> A supermarket or online delivery can save carbon, but buying less saves even more

Supermarkets stepped up their deliveries enormously during the COVID-19 lockdown and many customers have not gone back to their old ways. And the more people shop this way, the more efficient the deliveries become. In these sums I've assumed you live 1 mile from the supermarket, and that if it does the delivery, it combines yours with several other orders, making it half a mile of driving per drop.

It turns out that an electric delivery van comfortably beats driving in your own car, and even a diesel van is better. But a supermarket delivery isn't as good as you walking, cycling, or taking the bus or train to a supermarket or local store. And if you live out in the sticks and buy from a supermarket that is a long way from you and is therefore unable to combine deliveries, your own car could work out better, especially if you plan your trip and combine it with other reasons for going into town.

Similar considerations apply to Amazon or other online deliveries. If you can walk, bike, bus, or train to your local bookstore or main-street hardware store, keep doing it. You'll be supporting local business as well as keeping emissions low. An efficient bundled Amazon delivery saves on your making multiple shopping trips, but one trip for many items at the same time is probably more efficient than several unbundled home deliveries.

There is another catch with online shopping. Amazon makes it very easy to buy stuff and encourages us to buy too much. And all products have a carbon footprint. Buying less, using things for longer, and repairing them when they break are all key to a low-carbon life.

Traveling a mile by train

180 g CO_2e the subway
206 g CO_2e Amtrak Intercity standard class
270 g CO_2e commuter rail
412 g CO_2e Amtrak Intercity first class

> A 40-mile intercity train trip has the same footprint as nine pints of milk

Although trains can be a relatively green way to get around, the figures above show that emissions from traveling by train are higher than you might think. All the numbers provided include the direct emissions and electricity consumption of the moving train itself, but also try to account for the embodied emissions from train manufacture, the upkeep of the train tracks, and the running of all the infrastructure.[7]

The amount of energy required to propel a train down a track depends mainly on just a few factors:[8]

- **How fast the train goes** The air resistance goes up with the square of the speed.
- **How many stops there are** Each stop wastes energy, the exact amount being proportional to the square of the speed and the weight of the train. Some newer trains reduce this stoppage waste through "regenerative braking," similar to hybrid cars.
- **Rolling resistance of the wheels on the track** This is lower for trains than for cars because metal wheels on metal tracks are more efficient than rubber tires on asphalt. The resistance goes up proportionally with the weight of the train.
- **The type of fuel used** Electricity beats diesel because although there are inefficiencies in generating electricity from fossil fuels, the train engine can turn almost all of the power into movement and when it slows down, it can turn most of the kinetic energy back into electricity. A diesel

engine is much less efficient. In the UK, the carbon advantage of electric trains has increased noticeably as the grid has become greener.

Intercity trains go fast (that's bad), stop infrequently (good), but are also extremely heavy (bad). The weight of the train per passenger seat, amazingly, is around twice that of an average car. Professor Roger Kemp,[9] who has looked at this astonishing fact in detail, explains it in terms of overengineered safety—trains weigh at least twice what they need to because we have become obsessed with safety, even though train travel is one hundred times safer than driving. So perhaps as much as twice the energy is required to get our trains moving every time they leave a station.

First-class travel deserves mention because the number of seats you can squeeze into a first-class carriage is half that in standard class. This means that the weight being moved per person is doubled again; we're now up to the weight of four cars per seat. I sometimes board trains where half the length is near-empty first class and the rest is crowded standard class, suggesting that the real weight being hauled per first-class passenger may be even higher.

Interestingly, taking the subway generates slightly less carbon per passenger mile than intercity trains, despite stopping much more often. This is mainly because people are (or were in pre-COVID-19 times) packed in so tightly, nose to armpit. Other reasons are that the subway travels relatively slowly, is usually all electric, and has lighter trains.

Overall, trains are a lot greener (and a lot safer) than cars, though a sensibly designed gas-powered car can beat a train's carbon footprint if you fill it with people. Even two people traveling together are better off driving an efficient car than traveling first class. See also *New York City to Niagara Falls and back* (p. 124).

An hour watching TV

179 g CO_2e streamed on 13-inch MacBook Pro
206 g CO_2e broadcast TV viewed on 55-inch LED TV
279 g CO_2e streamed on a 55-inch LED TV
358 g CO_2e broadcast TV viewed on 42-inch plasma TV[10]

> The USA's favorite pursuit is relatively low carbon—even on a big screen

The average US adult spends a massive five hours a day in front of the box. Watching broadcast TV on a 55-inch LED screen would account for 376 kg (827 lbs) CO_2e per year. This is equivalent to driving 600 miles in an average gas-powered car in the US,[11] which is really quite a low figure for 1,825 hours of activity.

We also need to account for the emissions embodied in the TV set itself, from its manufacture and transport to the buyer, which for modern TVs turns out to be the biggest carbon component. A 42-inch plasma TV comes in at 580 kg (1,276 lbs), which means that after just one-and-a-half years, your total carbon footprint for the electricity would be bigger than the embodied carbon if you watched it for a US average number of hours per day. In contrast, a 55-inch LED screen is so much more efficient that it would take seven years for its electricity carbon footprint to equal its embodied footprint of 735 kg (1,617 lbs).

So, what are the carbon implications of trading in a plasma or CRT TV for an LED TV? Because modern LED TVs are more efficient for any given screen size than their older plasma or CRT counterparts, but have a larger embodied footprint, trading only makes sense if you watch a lot of TV. With the savings from switching to the more efficient LED TV, you might break even within three years if you watch the US average five hours a day. But if you watch only one hour a day, you should almost certainly stick with what you've got, or just possibly buy secondhand and pass your old one on. Watching with friends can clearly make things more efficient.

One addition is the footprint of transmission, which according to a study based on the BBC in the UK comes in at 8 g CO_2e per hour for digital broadcasting, 39 g CO_2e for satellite, and 42 g CO_2e for cable TV.[12] On top of that, you need to add about 11 g CO_2e for the embodied carbon of the set-top box.[13] And there is a shared carbon budget for creating the programs, though this is marginal in most cases (one report estimated 13.5 tons of CO_2 for an hour's worth of BBC TV content[14]), assuming it is shared by a reasonably large audience.

Video streaming can be better or worse than traditional broadcasting, depending on the network you stream and the viewing device itself. A modern laptop typically uses much less energy than a large TV and the embodied carbon also tends to be lower.[15] But if you stream onto a large TV screen, you forgo these potential savings. The carbon footprint of transmitting BBC iPlayer (similar to On Demand TV in the US) over the internet is comparable with satellite and cable TV, at 49 g CO_2e per hour.[16] BBC and Netflix are probably equally carbon intensive as they use the same data centers through Amazon Web Services. In addition, you have to add something for the embodied carbon of the Wi-Fi router. I assume it's about half that of a set-top box.[17]

Interestingly, unless you are using a mobile network, streaming a movie in high definition (HD) doesn't change the carbon footprint much compared to watching it in standard definition (SD). The critical factors are the time that you use the network for and how long your Wi-Fi or set-top box is turned on for rather than the rate of download. Mobile networks are generally more efficient than Wi-Fi unless you are streaming ultra-high definition (UHD).[18]

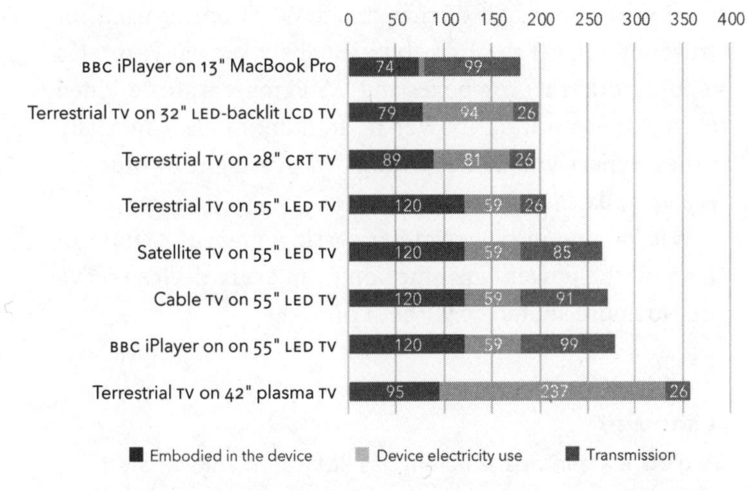

The carbon footprint of watching TV for one hour with different combinations of viewing devices and transmission options

Video, however, turns out to be one of the major drivers of global IT's rising footprint (see p. 187) because it is so ubiquitous—from Netflix, Amazon Prime, and YouTube to video conferencing and video ads embodied on many websites.[19] Our expectation to be able to stream from anywhere at any time of the day increases the need for carbon-intensive network infrastructure.

The largest part of the transmission is from network equipment in the home: the set-top box (for broadcast TV) because of their high standby power consumption, and the Wi-Fi router (for video streaming) because these devices are usually left on all the time. TVs typically use less than 0.5 watts in standby mode (that's much better than the 3 watts from ten years ago). So, a whole year of standby is just 1.2 kg (2.6 lbs) CO_2e of emissions.

Set-top boxes are a bigger deal, with active standby more likely to come in at 50 kg (110 lbs) CO_2e per year and passive standby at about half of that, and a Wi-Fi router used for streaming can get through 30 kg (66 lbs) CO_2e per year.[20] It's important that set-top boxes and Wi-Fi routers are designed to consume minimal power in standby mode, especially if they expect you to keep them on overnight in order to receive updates.

With a plug-in power meter costing about $25, you can check all the power consumption from every device you've got. No house should be without one.

A shower

55 g CO_2e 3 minutes, efficient gas hot water tank, aerated showerhead

270 g CO_2e 5 minutes in a 5 kW economy electric shower

430 g CO_2e 5 minutes in a typical 8 kW electric shower

1.8 kg (4 lbs) CO_2e 15 minutes in an 11 kW electric power shower

> Shorter and more efficient showers can save 350 kg (770 lbs) CO_2e a year–as much as a return flight from Seattle to Florida

Is three minutes enough time in the shower? It's about my average if I wake up half an hour before my train is due to leave. The time you spend showering, as you can see, makes a big difference.

The way you heat the water is also significant. In the US, nearly all showers are powered by gas, however in Europe and other parts of the world electric showers are quite popular. Gas is still a more carbon-efficient way of providing heat than electricity, provided you have a reasonably efficient hot water tank, although the gap will close as more renewable electricity becomes available. An aerated showerhead helps by making less water feel like more. It saves water and carbon without you having to forgo any comfort at all. If you are in a family of four and you each spend fifteen minutes in the shower every day, you may be able to reduce your household footprint by nearly a 400 kg CO_2e per year just by switching to an aerated showerhead. And you can cut those remaining emissions by *two-thirds* by having five-minute showers.

You will be swapping time in the shower for time doing almost anything else you want: reading a book, lying in bed, even both at once. And not just that. If you take all of these measures, your family could knock off 230 kg (506 lbs) CO_2e per year—and save nearly $550.

The showers in Iceland are worth a mention as the most luxurious I've ever had. Geothermally heated and almost carbon neutral, they are all the more enjoyable after a day out in Iceland's abundant rain and snow. Unfortunately, you have to fly to get there (see also *Taking a bath*, p. 82).

A unit of heat

30 g CO_2e solar water-heating panel
30 g CO_2e solar panel with a heat pump
90 g CO_2e solar panels and no heat pump
250 g CO_2e using a modern gas furnace
400 g CO_2e using an old gas furnace
650 g CO_2e using US grid electricity[21]
1.05 kg (2.3 lbs) CO_2e using Australian grid electricity

> A "unit" here means 1 kilowatt-hour (kWh), enough to run a "one-bar" electric fire for an hour, or to boil 4 gallons of water in a kettle

The most efficient means of domestic heat is solar water-heating panels. These have no operational emissions, so their carbon footprint is simply in the manufacture of the panel itself. This is generally pretty low compared to the energy they save. So, carbon-wise, they're fantastic.

Solar (photovoltaic or PV) panels have a higher embodied carbon footprint (see *Solar panels*, p. 154) but their electricity can be used to drive a heat pump. This works like a fridge in that it uses electricity to make energy flow from somewhere cool to somewhere warmer. It takes heat from the ground, or outside air, instead of from your food, and puts it into your home instead of into the pipes at the back of the fridge. In this way you can get about four times as much heat as would have been possible if you had just plugged your solar panel into an electric heater.

The problem with both solar heating and heat pumps is that they tend to deliver low-grade heat. In other words, they are all right for warming up baths and gently heating rooms, but not for boiling kettles or making toast.

In the middle of my scale is heat generated by an efficient gas-fired furnace, such as might power a new central heating system. In this scenario, your heating is done by fossil fuels, but at least you're using them fairly efficiently; the only losses will come from the energy disappearing out of the flue (typically around 10 percent).

The footprint of heating directly from grid electricity depends entirely on how your electric company sources its power. Coal is the worst, then oil, then gas; renewables are the best, along with nuclear energy. The precise footprint will depend on which country you are in (see A *unit of electricity*, p. 55), but with only a few exceptions the figure will always be high because most of the world's electricity is still generated from fossil fuels. And unlike with a gas furnace in your home, more than half the energy in the fuels is lost in the power station or transmission grid.

The UK has improved its electricity mix quite a bit since I wrote the first edition of *Bananas*, reducing its footprint 40 percent by phasing out coal and increasing use of renewables to 32 percent (but still only 5 percent of total UK energy).[22] In the US, however, despite some improvements, the footprint remains high as the electricity mix is still dominated by coal and gas. It is worth changing your electricity supplier to one that sources all its energy from renewables (this is not always as transparent as they suggest—see p. 55).

The advantage of electric heating is that it is easy to target it to just the places where you actually need it. In the low-carbon world, most of our heat will have to come from electricity.

A letter (and other mail)

280 g CO_2e 10 g letter on recycled paper (and then recycled by the recipient)

350 g CO_2e 25 g letter printed on virgin paper (and then sent to landfill)

2 kg (4.4 lbs) CO_2e a small catalogue or magazine (sent to landfill)

> If you get four letters and two catalogues or magazines a week, that's 280 kg (616 lbs) CO_2e per year–more than 5 percent of the 5-ton lifestyle

We tend to think that computers and phones have increased our carbon footprints. That's not entirely the case. If you have been browsing this book in order, you'll have noticed that the earliest and lowest-carbon entries include an email (a short message weighing in at 0.2–0.3 g) and a text message (0.8 g). By comparison, the old order of letters and packages is highly carbon intensive, most of its impact coming from the burden that it places on the infrastructure of our postal system: vans, trains, and sorting offices.

Mail clocks up a carbon footprint in four basic ways:

- **Paper production** The paper footprint depends on the recycled content and the efficiency of the mill. My estimates are based on paper that has a typical UK mix, with less than one-fifth recycled content. That gives it a footprint of 2.35 kg CO_2e per kilo (1.1 kg per pound). The best estimate for pure virgin paper is 2.59 kg per kilo (1.2 kg per pound), while fully recycled paper is about half (it takes half the energy to create new paper from old paper as it does to create paper from trees).[23]

- **Printing** To turn paper into glossy and enticing sales literature (or a magazine), I estimate an additional 350 g CO_2e per kilo (160 g per pound).
- **Postage** For a standard letter, this accounts for most of the footprint. It's hard to account for but if you take the footprint of the postal services sector as a whole and divide it by the turnover of that sector, you can get a broad idea of the carbon footprint per unit of cost. In the US, it comes to about 300 g CO_2e per dollar spent.
- **Decomposition** A good deal of junk mail ends up in landfill, where it decomposes anaerobically and produces methane. For this, I have allowed 1 kg CO_2e per kg of paper (450 g per pound).[24] You can avoid this, of course, by recycling as much mail as possible. This is okay to do even if the letter has a plastic window. But do remove any other plastic, such as film wrap.

What can you do to cut down? Eliminating junk mail is the biggest step. It will declutter your life as well as saving carbon. To avoid it, use a free online junk-mail opt-out service (see endnote).[25] To deal with the ones that still get through, keep a stack of printed labels saying "Return to sender. Please remove us from your database."

Finally, a message to the instigators of junk mail: people will think badly of you for using high-carbon marketing

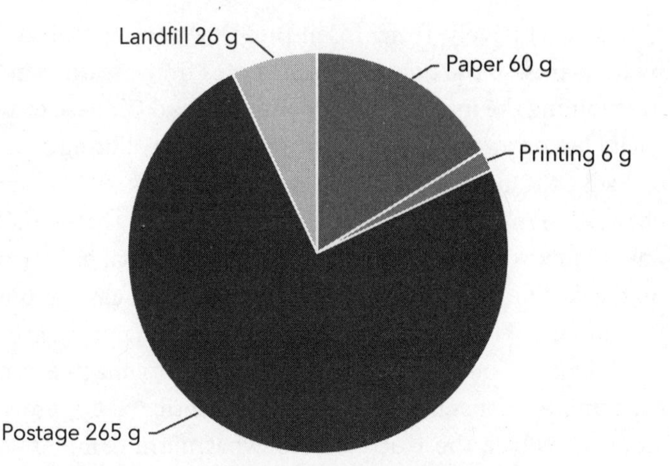

The carbon footprint of a 25 g letter, printed on virgin paper, sent by ground mail, and thrown into landfill

techniques. If you must use mailers, at least keep databases up to date, use recycled paper, and keep your messages short and light.

A unit of electricity

9 g CO_2e from the Icelandic grid
340 g CO_2e from the UK grid
650 g CO_2e from the US grid
980 g CO_2e from the Chinese grid
1.05 kg (2.3 lbs) CO_2e from the Australian grid

> The carbon impact of using an additional unit of electricity is higher than we tend to think

Electricity generation is one of the principal causes of global carbon emissions. However, the carbon cost of each unit of power varies greatly, depending on the precise mix of generating fuels used in your country. Icelandic electricity comes almost exclusively from fossil-fuel-free geothermal and hydro plants, so the only footprint comes from creating and maintaining the infrastructure. Australian and Chinese electricity, by contrast, comes mainly from highly polluting coal.

Sadly, the overall footprint per unit in the US has not changed significantly in the last ten years, as the overall energy mix still relies heavily on fossil fuels. Coal has been on the decline, replaced by an increase in natural gas use, but the amount of renewable energy in the mix remains small.

Thinking of our carbon footprint, we tend to imagine that each unit we consume causes a fixed quantity of CO_2 emissions. However, the truth is somewhat more complex. A more meaningful way to think about the carbon footprint of your electricity use is to think of it as being *additional* to all the power consumption that was already going on before you flicked on the light or appliance.

Look at it this way. The extra demand that you place on the grid is met entirely through *additional* fossil fuels, because the renewables in your country will already be running at full capacity. In other words, when you turn the lights on, you don't personally affect the amount generated by renewables because they are already going flat out. Rather, what you trigger is almost certain to be a lump of coal thrown into a power station or a puff of gas going into a turbine. This is the case even in countries where all electricity comes from renewable energy sources or nuclear power, because adding to demand reduces the amount of electricity that those countries are able to export, thereby increasing fossil-fuel generation in other nations.

If you look at it in terms of marginal demand (see the table opposite), each unit of electricity you consume has a footprint of around 400 g CO_2e per unit if it's generated by gas or around 1 kg if it's derived from coal.

Country	Direct emissions from power generation, per unit consumed	Estimate of total footprint per unit consumed	Estimated footprint of marginal demand
Australia	0.86	1.05	1.06
China	0.80	0.98	1.06
Iceland	0.00	0.01	0.01
UK	0.28	0.34	0.4
USA	0.53	0.64	1.06

Three ways of looking at carbon intensity in kg CO_2e per kWh

One exception is Iceland, where, for the moment at least, it looks as though you can more or less use as much electricity as you like without boosting your footprint. The country is overflowing with hydroelectric and geothermal power; indeed, you can see the energy almost everywhere you go, boiling out of the mud and pouring over waterfalls. There are no direct emissions from electricity production, but I have guesstimated that there is a small amount for the embodied carbon in the infrastructure.

Of course, once Iceland works out how to export its clean energy (or use it for other means[26]), or how to import enough of the world's heavy industry to use up its capacity, electricity will become a scarce resource for Icelanders, just as it is for the rest of us.

A newspaper

210 g CO_2e *USA Today* (weekday), recycled
330 g CO_2e *New York Times* (weekday), recycled
820 g CO_2e *New York Times* (weekday), sent to landfill
1.9 kg (4.2 lbs) CO_2e weekend paper with supplements, recycled
4.7 kg (10.3 lbs) CO_2e weekend paper sent to landfill

> Two big weekend papers a week adds up to 200 kg (440 lbs) CO_2e a year, even if you recycle them–equal to flying from Atlanta to Fort Lauderdale in Florida

It's amazing how energy-hungry newspaper production can be. And the figures provided here are on the low side because none of them take account of the footprint of journalism itself including the newspaper offices and staff flights.

The carbon footprint of a weekend newspaper. Sending paper to landfill causes methane emissions and means that more carbon-intensive virgin paper has to be produced.

A single weekend paper (typically 1.25 kg, or 2.75 lbs) could add up to 5 percent of a 5-ton lifestyle if you don't recycle it. Opting for a slimmed-down weekly paper, such as USA *Today*, is one good way to reduce emissions. Another is to get your news online, which comes in at around 1 g CO_2e per minute from a smartphone (a little more on an iPad) and wins hands down, unless you spend at least three hours reading it through.

Recycling papers is important for two reasons. First, if paper is disposed of in landfills, it emits methane as it rots. Second, for each newspaper that isn't recycled, one more newspaper's worth of virgin paper has to be manufactured. So, throwing your paper in the general waste more than doubles its footprint.[27]

A liter (32 oz) bottle of water

320 g CO_2e locally sourced and distributed
400 g CO_2e average
480 g CO_2e transported 600 miles by road

> An avoidable disaster that keeps on growing

In the ten years since the first edition of this book, worldwide bottled water consumption has almost doubled from 52 to 103 billion gallons per year.[28] This means bottled water alone accounts for over a quarter of a percent of the world's carbon footprint. It is 1,000 times more carbon intensive than water from the tap. So, for anyone living in a country where the tap water is safe to drink, knocking the plastic bottles out of our lives has got to be a simple win.

The carbon comes mainly from packaging and transport. Processing water, whether still or carbonated, has a marginal carbon impact. But there is an add-on of 83 g CO_2e per quart just for the plastic, on top of which is a further 20 g CO_2e to melt the PET (polyethylene terephthalate) balls down and mold them into bottles. Transport is also significant, because water is so heavy. If the bottles travel 600 miles by road, that can add a further 80 g CO_2e per bottle.[29]

Why don't all train stations and town centers have drinking fountains?[30] Other than during a pandemic, it would be a simple win for everyone, except those who make money selling the bottled stuff. And that is part of our answer. If we got rid of bottled water, people would be financially better off, but the economy would look as though it had slowed down a fraction. This is a nice illustration of how inadequate it is to measure how we are doing solely in terms of our economic growth. The economy will actually recede as we all get better off.

Sadly, in the US, restaurants and cafés are not required to provide tap water free to customers, although most will if asked—and you should never be embarrassed to ask.

A bowl of oatmeal

110 g CO_2e traditional Scottish (made with water)
450 g CO_2e made with soy milk
450 g CO_2e made with half cow's milk, half water
800 g CO_2e made with cow's milk

> It all depends on the milk

Oats are a fantastic low-carbon food that also happens to be healthy and tasty. If you fed yourself entirely on oatmeal made with cow's milk, your food footprint would be 1.6 tons CO_2e per year. If you stuck to the Scottish water-based version, you could get by on about half a ton. Soy milk (I can't bring myself to call it "soy drink" as the producers are legally obliged to do[31]) offers a nutritious and perhaps even tastier low-carbon alternative to cow's milk. Oat milk is another option, if a bit conceptually weird paired with cooked oatmeal.

Assuming a more varied diet, a bowl of half-milk oatmeal every day would be about 3 percent of a 5-ton lifestyle. I have assumed typical US cow's milk, which has a somewhat lower carbon footprint than the global average (see A *pint (16 oz) of milk*, p. 83). Sprinkling 10 g of sugar on top would add another 13 g CO_2e.

The cooking is about a third of the footprint of the traditional Scottish version. I've assumed that you cook it on the stove and never have the lid on, because you are stirring like crazy, trying to save yourself a dishwashing nightmare. Better still, the microwave is lower carbon than an electric or gas hot plate and doesn't cause sticking; but keep a close watch or it will turn into an exploding mess. As my family will tell you, I am the last person who should be writing a cookbook.

Oatmeal with cow's milk	**Footprint (g CO_2e)**	**Calories**	**Calories per kg CO_2e**
Oats (50 g)	70	185	
Fully cream milk (350 ml)	0.80	0.98	
Bring to the boil stirring carefully	0.00	0.01	
TOTAL	797	418	530

Traditional Scottish oatmeal	**Footprint (g CO_2e)**	**Calories**	**Calories per kg CO_2e**
Oats (50 g)	70	185	
Water (350 ml)	0	0	
Bring to the boil stirring carefully	40	0	
TOTAL	110	185	1,681

A roll of toilet paper

450 g CO_2e recycled paper

730 g CO_2e virgin paper

> This can account for 84 kg (185 lbs) CO_2e per year–about 1.7 percent of the 5-ton lifestyle

The typical North American goes through 26 kg (57 lbs) of toilet paper per year, compared to 16 kg (35 lbs) per year for a Western European and less than 1 kg (2.2 lbs) per year for an African or East Asian. The global average is 5.2 kg (11.4 lbs) per person per year, totaling about 40 million tons and accounting for 0.25 percent of annual emissions.[32]

I'm not sure I want to launch into a detailed exploration of bathroom technique here, but 1.7 percent of the 5-ton lifestyle seems so high for such a simple and brief part of our lives that it is worth a moment's personal reflection. My numbers show that a sense of economy is in order. If, as I suspect, many of us could halve our usage without any negative side effects, it could be an easy and worthwhile carbon win. Perhaps all those ads trying to persuade us that that an over-pampered bum is an essential feature of a good life aren't the last word.

My footprint figures here are based on numbers from Tesco, whose research suggests a carbon cost of 1.1 g per sheet for their recycled stuff and 1.8 g for traditional paper.[33]

Washing dishes

0 CO_2e by hand in cold water (but the plates don't get clean)
360 g CO_2e by hand, sparingly, in warm water[34]
800 g CO_2e in a dishwasher, eco-setting (50°C/120°F)
1 kg (2.2 lbs) CO_2e in a dishwasher, heavy duty setting (65°C/150°F)
3 kg (6.6 lbs) CO_2e by hand, with extravagant use of water

> Running a dishwasher four times a week is equivalent to driving an average car 264 miles

Dishwasher versus hand washing plates is a hot debate among us carbon footprint geeks. So, here are the results:

The most careful hot-water (but not too hot) hand washing can just about beat the dishwasher on carbon. But it loses out badly on both hygiene (nearly 400 times the bacteria count[35]) and time (four times as long). Overall, the dishwasher (based on a Bosch with A+ efficiency rating) wins, even if you add in its embodied emissions. And that's even without the carbon saving possible if you set your machine to run in the middle of the night when electricity demand is low and the grid becomes more efficient (see *A unit of electricity*, p. 55).

My figures for the dishwasher are based on always running a full load and include 100 g CO_2e for the wear and tear on the machine (based on a fairly expensive built-to-last model that you keep for twelve-and-a-half years—as with all appliances, the longer you keep it, the lower the embodied carbon per use).[36]

The conclusion, then, is use a high-efficiency modern dishwasher. It simultaneously helps the planet, your health, and your lifestyle. When you buy one, choose a make that will last, and look after it. Try to always run it full, use the economy setting, and, if you can manage the logistics, run it in the middle of the night.

I haven't, incidentally, included anything for detergent or water consumption, because they are nothing compared with the impact of heating the water.[37]

Oh, and a final note: I have known people who routinely wash their stuff by hand before putting it in the dishwasher. This must be the worst of all options and ranks alongside ironing your spouse's socks. If this is your routine, liberate yourself!

A 250 g (8 oz) clamshell of strawberries

490 g CO_2e grown in season, locally

770 g CO_2e frozen

1.7 kg (7.74lbs) CO_2e flown in out of season from Mexico to New York

> Out-of-season strawberries have a footprint more than three times higher than tastier seasonal ones

The estimate above for a (250 g/8 oz) clamshell of local, seasonal strawberries varies depending on such things as the soil and the use of fertilizers and polytunnels.[38] Some of these variables increase both the yield and the emissions per acre, so whether they result in more or less carbon per strawberry is quite complex. But the numbers are all so much better for local, seasonal strawberries than for the out-of-season versions that a good enough rule of thumb is just to stick to those grown in your own country, unless your government subsidizes the heating of greenhouses, as in the Netherlands. This kind of hothousing is, broadly speaking, just as bad as airfreighting the fruit from hotter countries (see *Flying from Los Angeles to Barcelona return*, p. 152, and A 250 g (8 oz) *bunch of asparagus*, p. 86).

So, the best advice is to wait until strawberries are in season, then enjoy them twice as much. Or, if you can't wait, try frozen ones. For what turns out to be a fairly modest carbon price, all the taste and nutrients locked down within hours of harvest emerge in perfect shape in your kitchen at the exact time of your choosing. This also makes it easier for you to cut waste. The same principles apply to all fruit and vegetables. Canned fruit lies somewhere in the middle of the range, in carbon terms, along with those traveling moderate distances by road and boat from warmer climes.

All the figures here have taken account of the 23 percent average wastage between field and checkout. A small amount of the footprint is the packaging, which is actually in a good cause if it enables more of the strawberries to find their way into our mouths. The footprint of the plastic will typically be lower than that of the wasted fruit, although plastic presents its own problems (see 1 kg (2.2 *lbs*) *of plastic*, p. 106).

500 grams to 1 kilo (1.1 to 2.2 pounds)

An ice cream or popsicle

70 g CO_2e an 80 g popsicle from the supermarket
140 g CO_2e an 80 g ice cream bar
500 g CO_2e a big dairy chocolate chip ice cream from an ice cream truck

I throw my hands in the air—the figures above are back-of-the-envelope guesstimates. But they come from a broad understanding of the footprints of different food ingredients, transportation impacts, and mobile refrigeration.

So, a popsicle is essentially frozen sugary water, and in a supermarket the refrigeration is likely to be relatively efficient. If you buy a box and store them in your own freezer, that will increase the footprint a little. As would an electric ice cream truck.

A traditional ice cream truck's footprint is much higher for three reasons: it's bigger, the ice cream is dairy based (with the associated footprint from the dairy industry), and it's kept cold in a less efficient mobile refrigeration unit, often belching out diesel fumes.

Chocolate chips make the footprint of an ice cream even higher, but I think we can all agree it's worth it.

Driving a mile

260 g CO_2e in a mid-sized five-door electric car
290 g CO_2e in a small car doing a steady 60 mph (100 kph)
630 g CO_2e in an average US car at 25 miles per gallon
1.26 kg (2.7 lbs) CO_2e in a Range Rover Sport, new but not looked after, doing 90 mph (140 kph)

> Driving a car the US average annual distance of 13,500 miles can use from 70 to 340 percent of the 5-ton lifestyle

My numbers are higher than those you normally see for driving. That is partly because I am including the emissions from the extraction, refining, and transportation of fuel as well as the burning of it. Even more importantly, I am factoring in the manufacture and maintenance of the vehicle itself.

As a rule of thumb, about half of the carbon impact of car travel comes out of the exhaust pipe.[1] A few percent come from the fuel before it is burned (see *A 50-liter (13-gallon) tank of gas*, p. 129). The rest, typically 40 percent of the footprint, is associated with the manufacture and maintenance of the car. Big, expensive, new cars have more of their embodied emissions attributable to each mile of driving. An older car that is still fairly efficient could beat a new efficient vehicle by virtue of having had its embodied footprint written off (see *A new car*, p. 157).

The electric car is a clear winner, but still has a substantial footprint because of the embodied carbon and the footprint of the electricity.

Among gas-powered cars, the low end of the scale above is calculated for a well-maintained low-emission vehicle (such as a Fiat 500 or Ford Fiesta) traveling at a steady 60 mph. My own Peugeot 107 can do 75 miles to the gallon under these conditions. With four people, the carbon comes out at 73 g CO_2e per person per mile.

At the high end of the scale, we have a single person in an SUV that looks more like a tank, cruising at 90 mph or driving aggressively in urban conditions. In these conditions, a vehicle of this type may achieve as little as 18 miles per gallon.

The choice of a new car is crucial to your carbon footprint. Vehicle emissions have not decreased at anything like the scale we might have hoped for ten years ago, and the reason for that is too many people buy SUVs rather than fuel-efficient cars. This is a disastrous fashion: big SUVs are totally unnecessary for most people. Any supposed safety advantage in a crash comes at the expense of whoever you crash into. They should be taxed very heavily indeed. If you are replacing a car, take the plunge and go electric (and not bigger than you need); you'll reduce your maintenance and running costs enormously, as well as your emissions. Almost nobody who goes electric regrets doing so.

But it's not just what model you drive that matters. Here are nine good ways to reduce the carbon footprint of your car use:

- Put more people in the car or even join a ride-sharing service. A car can compete in emissions with train travel if you are avoiding four separate journeys. Typical saving: 50–80 percent.
- Look after your car so that it will do at least 200,000 miles in its lifetime and runs as efficiently as it can. Typical saving: 30 percent compared with the average (see A *new car*, p. 157).
- Accelerate and decelerate gently, avoiding braking as much as possible. Typical saving: up to 20 percent in urban conditions.
- Drive at 60 mph on highways. Typical saving: 15 percent compared with 70 mph.

- Keep the windows up when driving fast, and the air conditioning off. Typical saving: 2 percent.
- Keep the tires at the right pressure. Typical saving: 1 percent.[2]
- Avoid rush hour (see A *rush-hour car commute*, p. 114).
- Drive safely (see A *car crash*, p. 170).
- Use the train, bus, or bike if traveling alone. Typical saving: 40–98 percent (see *New York City to Niagara Falls and back*, p. 124).

A latte (or a tea or coffee)

22 g CO_2e black tea
47 g CO_2e tea with soy milk
49 g CO_2e black coffee, instant
71 g CO_2e tea with cow's milk
87 g CO_2e black coffee (drip, Americano, or filter)
288 g CO_2e large oat milk latte
308 g CO_2e large soy milk latte
552 g CO_2e large cow's milk latte
+ 110 g CO_2e for a typical disposable cup[3]

> A cup of black tea is low carbon ... but a cow's milk latte every day could make up 4 percent of the 5-ton lifestyle

The shock in the figures above is the milk. If you take your tea with milk, and you boil only the water you need, then the milk accounts for three-quarters of the total footprint (see *A pint (16 oz) of milk*, p. 83). The obvious way to slash the footprint of your tea is reduce the amount of milk, switch to soy or oat milk, or drink your tea without milk.

Coffee also has a significant footprint,[4] much higher per cup than tea. Instant has a lower footprint than ground

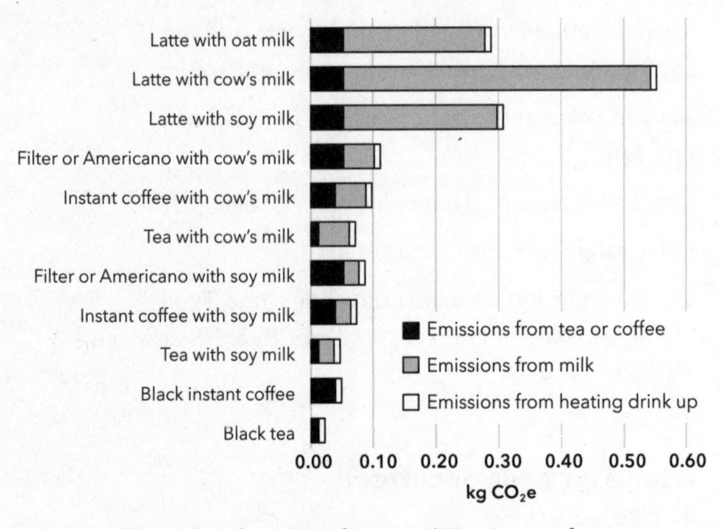

The carbon footprint of a 250 ml (8 oz) mug of tea or coffee. Add 110 g CO_2e for a disposable cup.

because a factory is more efficient at extracting the coffee than the device you use to make the coffee from ground beans, regardless of whether you go for filter, drip, or espresso. Admittedly, though, instant coffee doesn't taste good.

I have based my latte sums on the large kind that some of the coffeehouse chains encourage you to quaff. For the same impact, you could have nine carefully made Americanos or filter coffees, or twenty-five black teas. The large lattes also mean you are drinking an extra half a pint of milk, perhaps without realizing it.

Ten years ago, at my work, we tried and failed doing without milk in our drinks. More recently we have learned to love oat and soy milk. Barista oat milk is better in coffee, as it doesn't curdle.

In all the sums above, I have assumed that you only boil as much water as you need.[5] If you boil excessive water (as most people do), you could easily add 10 g CO_2e to your drink. Boiling more water than you need wastes time, money, and carbon; if you haven't yet developed perfect judgment, simply measure the water with a mug.

For coffee on the go, add a totally unnecessary 110 g for a typical disposable cup or the same if you keep losing your reusable one. The push for reusable mugs is, on its own, a trivial step in the push toward the low-carbon world. This is reassuring during the pandemic switch back to single-use cups. But it will be good to get back to reusables once the pandemic is over, if only because it helps us to become more carbon conscious. I'd like to see mugs with a caption saying, "Every time I use this cup, it reminds me to care even more about the bigger carbon issues in my life." Finally, think about your mugs. Buy sturdy ones, look after them, and save hot water by only washing them up at the end of the day, rather than using a fresh mug for every cup.

1 kg (2.2 lbs) of trash to landfill

590 g CO_2e average trash can contents
9.1 kg (20 lbs) CO_2e aluminum and copper

> The average American sends 500 kg (1,100 lbs) of trash to landfill or incineration[6] each year and recycles 260 kg (572 lbs), or about 6 percent of a 5-ton lifestyle

In the US, average emissions from trash cause 290 kg (638 lbs) CO_2e per person each year. By trash, I mean things we dispose of by putting them in a trash can, as opposed to recycling or composting them. Looked at this way, there are two parts to the footprint.

First, there are the landfill emissions, which are due mainly to stuff decomposing underground. This anaerobic decomposition produces methane, most of which gets captured in a well-managed landfill site, but at least some of which escapes to warm the world. (This isn't an issue for

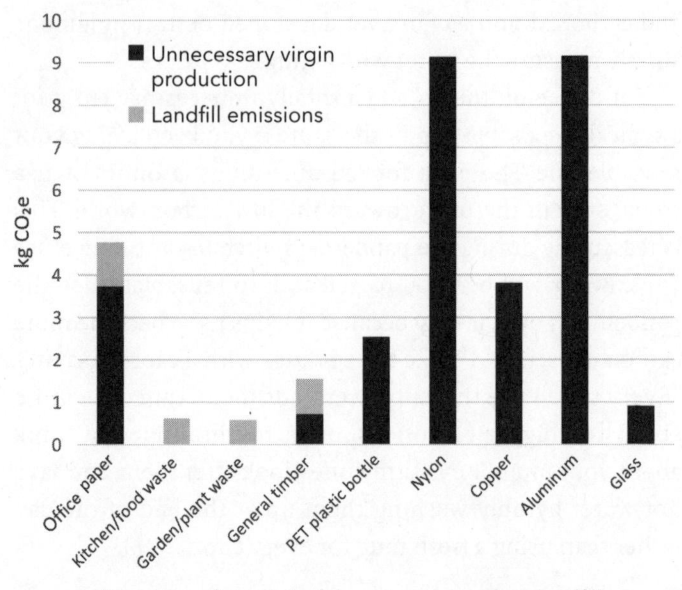

The carbon footprint of throwing out trash to landfill

metals, glass, plastic, and many building materials, of course, because they don't decompose in the way that food, paper, and garden waste do.) There is also a little bit of fossil fuel required to run a landfill site.

Second, there is the fact that, by not recycling something, you are forcing more virgin materials to be used in future products. This isn't an issue for food, for which recycling was never an option. But for metals, plastics, and paper it is a big deal.

The diagram above shows the impact of 1 kg (2.2 lbs) of various items you might throw into the trash can and from there to the landfill. As you can see, if that item is aluminum or nylon, it is particularly important that you recycle it.

The significance of food waste is underplayed in the diagram as it shows only the difference between landfill and recycling or composting (not for the original and wasted food production).

A loaf of bread

630 g CO_2e an 800 g loaf produced and sold locally[7]

1 kg (2.2 lbs) CO_2e same 800 g loaf transported extensively by road

> Bread is good: a year's caloric intake can be had for just over a quarter of a ton CO_2e

As the pie chart below shows, more than 60 percent of the emissions from a loaf of bread come from the actual wheat cultivation (and most of that is just the fertilizer use). Nearly a third comes from the milling and the baking. Transportation is only one-twentieth if it is bought locally, although this jumps up to 40 percent if it's transported a long way by road. The bag is a very small consideration and, if it helps to keep the bread fresh for longer, it is probably well worth it.

So, bread is a great low-carbon food provided we actually eat it. There's the catch. It gets thrown away too often, because we are fussy eaters and because it doesn't keep well.

Tristram Stuart's eye-opening book *Waste* has a picture of a Marks & Spencer sandwich factory systematically discarding four slices from every loaf: the crust and the next slice from each end.[8] The remaining slices get made into fresh

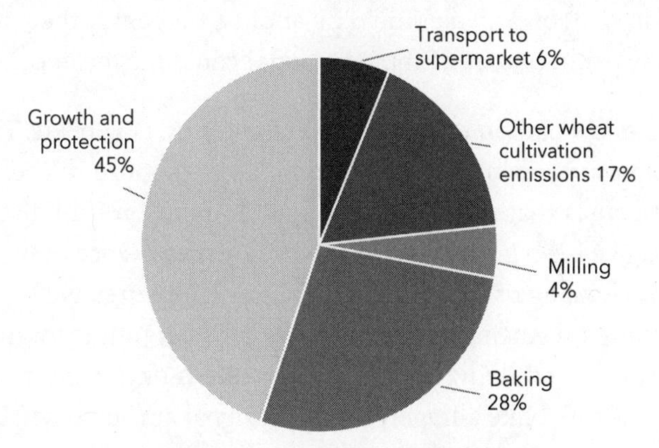

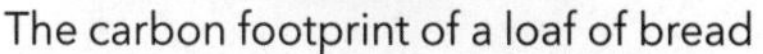
The carbon footprint of a loaf of bread

sandwiches but are still at risk of being thrown out before they are sold. Only once it is safely through the checkout do the odds of a sandwich being eaten start looking good, but there are still such hurdles as over-catered corporate lunches.

Loaves sold straight to consumers are no better because the shelf life is so low. Plenty is thrown out by the supermarkets and plenty more goes stale or ends up in a half-eaten sandwich. To keep the carbon cost of your bread to a minimum, buy only what you need, enjoy the crusts, and get your children to do the same. Remember that bread freezes well if you realize you have too much, and find uses for stale bread: toast, dunked in soup, and so on.

A pint (16 oz) of beer

365 g CO_2e and falling fast–BrewDog beer from the keg
420 g CO_2e and falling fast–BrewDog beer from the can
540 g CO_2e average locally brewed cask ale at a pub[9]
650 g CO_2e average local bottled beer from the store
760 g CO_2e foreign beer at the pub
830 g CO_2e average long-distance bottled beer from the store

> A daily pint of draft ale would be about 4 percent of the 5-ton lifestyle, but bottles of imported lager can triple the impact

At the low carbon end of my list come BrewDog beers. This company is making such an exemplary response to the climate crisis that they deserve a special mention (full disclosure: I've been helping them on that journey). Once they put in place many of the standard efficiency measures, we began working on cutting a further 40 percent per pint from their footprint, and doing so within a twelve-month timeframe. That seemed like a hugely ambitious goal at the outset, but

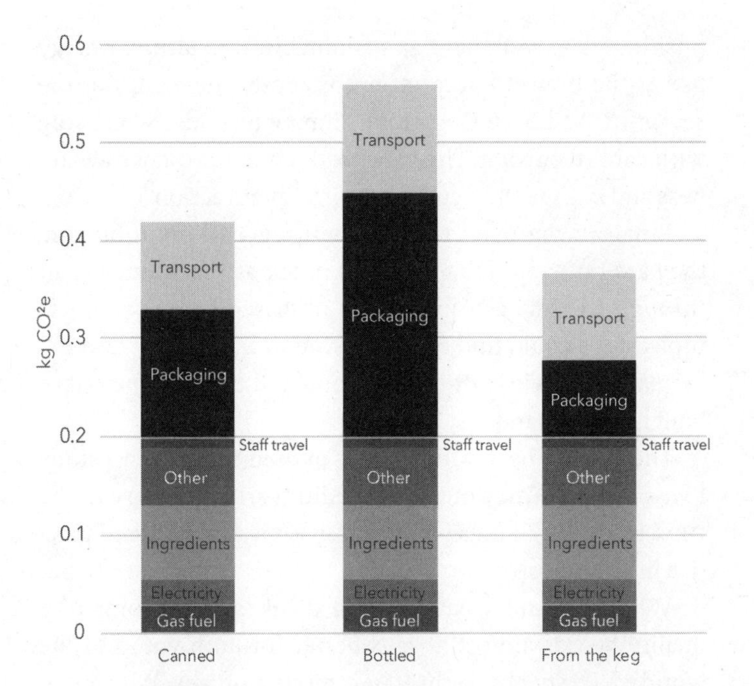

The footprint of a BrewDog beer per US pint. Cans are better than bottles, and straight from the keg is better still.

we found some surprisingly easy wins. For a start, switching from bottles to cans saved over 100 g per pint on packaging. Because cans are lighter, this cut the transportation footprint, too. Even after shifting to cans, packaging still made up about 30 percent of the footprint. Making sure their can supplier uses 100 percent recycled aluminum or, better still, used some of the billion or so leftover cans that other drink manufacturers throw away every year contributed further big savings. Happily, I'm told, cans are better for the quality of the beer, allowing in no light and always having a perfect seal. At the start, transportation made up over a fifth of the footprint with potential savings coming from going electric and optimizing logistics. Ingredients made up almost another fifth, pushing us to look hard at sustainable agriculture. Energy use at the brewery accounted for about one-eighth of the footprint. And BrewDog's total climate response won't stop with carbon cutting. They are working hard to raise awareness and engage their customers in climate action.

Finally, whatever their remaining carbon footprint, they are now removing twice as much as that from the air through the highest quality nature-based carbon removal projects we could find for them. And in the future, they will be planting their own native woodland on land they have bought in Scotland.

The average beer numbers I've quoted come from a study I was asked to carry out for a medium-sized brewery in the UK; I think the figures will be fairly typical of standard practice in the industry.

Wherever and whatever you drink, a single pint of a quality beer is almost always better for both you, and the world, than spending the same money on several cans of bargain-basement brew.

A 10-inch pizza

1 kg (2.2 lbs) CO_2e Mighty Veg (vegan, no cheese)
1.4 kg (3.1 lbs) CO_2e Margherita (cheese and tomato)
2.2 kg (4.8 lbs) CO_2e pepperoni (with cheese)
2.2 kg (4.8 lbs) CO_2e quattro formaggi
3.7 kg (8.1 lbs) CO_2e meat lover's[10]
+ 130 g CO_2e for a takeout box
+ 340 g CO_2e per mile for delivery on a scooter
+ 630 g CO_2e for delivery by car

> Basic pizzas are low-carbon—it's the cheese and toppings that count

More or less any meal based on bread gets off to a good start in terms of carbon: sandwiches, Indian food with chapatis,

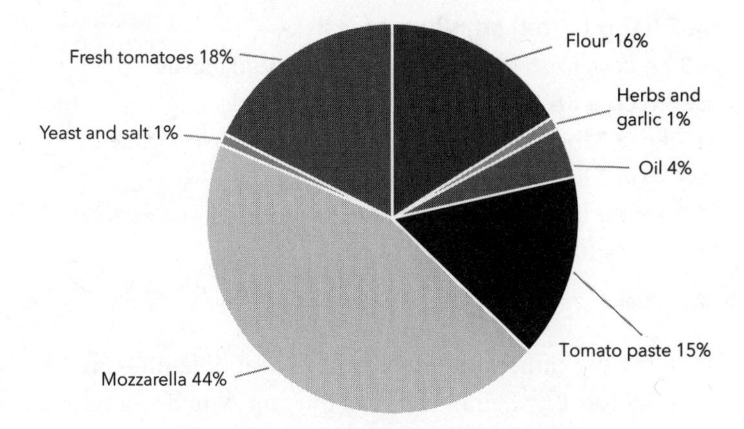

The carbon footprint of a Margherita pizza

and, perhaps best of all, pizza. The pizza base has a footprint of only 240 g CO_2e. It's what you put on top that counts.

A spread of tomato paste and fresh tomatoes add about 460 g CO_2e. Our vegan option above adds another 300 g CO_2e worth of onions, sweetcorn, peppers, and broccoli, oil, and herbs. Add cheese and you are soon racking up the footprint—60 g of mozzarella for a Margherita adds 600 g CO_2e, while a quattro formaggi with 95 g of cheese adds more than 1 kg (2.2 lbs) of CO_2e. But that's nothing compared to a meat lover's pizza. Piling 30 g each of ham, beef, and bacon on top of a Margherita gives it four times the vegan pizza's footprint.

Delivery on a cheap scooter adds 340 g per mile, or 630 g per mile in an average US car, and the box adds another 130 g (not so bad if you keep it clean enough to be recycled or if your municipality collects food-soiled paper products with the compost). But, as we all know, pizza is best eaten within minutes of being baked.

A 200 g (7 oz) serving of fish

480 g CO_2e fresh mackerel, caught and sold locally[11]

880 g CO_2e fresh cod, caught and sold locally

1.7 kg (3.7 lbs) CO_2e fresh trout, caught and sold locally

1.9 kg (4.2 lbs) CO_2e tinned tuna

4.5 kg (9.9 lbs) CO_2e fresh lobster, transported extensively by road

> As usual, local is best

Fish is a big carbon improvement on beef or lamb—and it's healthy, too. But, before you rush out and switch to a full-on pescatarian diet, bear in mind that many of our fish stocks are getting dangerously depleted and if we all switched over from ruminant meat we'd wipe out global fish stocks in decades. In addition, much of the world's sea fishing is done in a hideously unsustainable way, while farmed fish incur all the usual problems and inefficiencies of animal farming.[12]

I hesitate to mention that, in the studies I looked at, overfished cod comes out with a slightly lower carbon footprint than salmon. Once again, we each have to balance up carbon emissions with other concerns. For the sustainability rating of several types of seafood, check out the Marine Conservation Society's handy guide at fishonline.org.

When canned fish are compared with fresh fish, the refrigeration is a bigger deal than the can, so fresh fish come out slightly higher in carbon. But then the fresh version is 100 percent fish, with no added oil, and it usually tastes better, too. Fresh fish has a similar footprint to frozen as long as it has traveled to market on a boat.

Two to avoid are airfreighted fresh tuna or other fish, and lobsters, which are often transported long distance. Imported king prawns are even worse, particularly if farmed in South Asia, which often destroys mangrove plantations, the rainforests of the ocean.

Spending $1

166 kg (365 lbs) CO_2e on a rainforest restoration project[13]
1.9 kg (4.2 lbs) CO_2e on solar panels[14]
0.8 kg CO_2e on a typical supermarket shop[15]
3.2 kg (7 lbs) CO_2e on a mile's worth of gas for your car[16]
5 kg (11 lbs) CO_2e on the electricity bill[17]
5.6 kg (12.3 lbs) CO_2e on a budget flight

> With wealth comes carbon responsibility

Unless you are spending on something that reduces emissions—in which case, solar panels could be the best carbon investment you ever make—just about anything you choose to spend money on will increase your carbon footprint. Ripples of economic activity create ripples of emissions. So, what are you going to do with your $1?

Alas, the worst of all carbon investments is travel, especially if you fly off on holiday. Among the wealthier sectors of the US, carbon footprints are dominated by emissions from flights. It will be repeated many times in this book, but you can use up a full 5-ton carbon budget by a couple of long-haul flights. Even a return economy flight from Los Angeles to Chicago accounts for 1 ton of CO_2e.

On a far lower level, leaving the lights on unnecessarily is one of the cheaper ways of trashing the planet. As is using more gas than you need. Or buying carbon-intensive food at the supermarket. We need to be carbon aware of all our activities as consumers. Indeed, the less we consume, the lower our carbon footprint. It's almost as simple as that. Buying stuff has a carbon price.

Investments, however, can create *negative emissions*. On the domestic front, we should all look at the possibility of installing solar panels. And, if we have spare money, we should invest in companies or give to charities that reduce the world's footprint. We can do meaningful personal offsets by helping preserve rainforests, or reforesting, or boosting solar capacity in the developing world. And an investment in a wind farm or solar energy company will be using your wealth to bring about a low-carbon world. For more on all this, see the *What can we do?* section at the end of this book (p. 209).

A paperback book

400 g CO_2e recycled paper (if every copy printed is sold)
1 kg (2.2 lbs) CO_2e regular paperback book (60 percent sold)
2 kg (4.4 lbs) CO_2e the same book on thick virgin paper, with half the copies getting pulped

> The carbon footprint of a typical paperback is about the same as watching six hours of TV

Reading is a low-carbon activity and there is plenty of room for it in the sustainable lifestyle.[18] Why? It's hard to drive or shop while you read, so a gripping book halts the consumerist lifestyle in its tracks. And, if it's good, you'll hold on to it (sequestering the carbon on a fairly long-term basis), pass it on to a friend, or take it to a secondhand store.

My figure for a regular paperback is based on a 250 g book printed on paper from a typical mix of virgin and recycled pulp. I've assumed that 60 percent of all copies made are actually sold, though I've heard more pessimistic estimates than this. The economies of scale in printing are such that it pays to print too many. At the high end, the same book is printed on heavyweight, high-gloss virgin paper and weighs 350 g. Half of the print run never leaves the warehouse due to lack of orders and is pulped without ever hitting the shelves, let alone readers.

At the low end, the book still weighs 250 g but is printed entirely on recycled paper. Roughly speaking, it takes about twice as much energy to make paper from trees as it does from recycled pulp, though this varies enormously depending on the efficiency of the paper mill and the quality of the paper.[19]

I'm guessing that, in carbon terms at least, you are currently holding a better-than-average paperback, because my publisher thinks about these things. However, once you stop to think about it there are all sorts of difficult questions about what to include in the sums. I haven't included the electricity burned by my computer as I typed, or the editor's, proofreader's, indexer's, designer's work, or any part of the footprint of my publisher's offices at Greystone, or a host of other possible elements. Nonetheless, I hope that some of the ideas I've put forward in this book mean that it will pay for itself in carbon terms fairly easily. You only have to cut out about 2 car miles to cancel out its production.

Kindles and other e-book readers deserve a mention. I guesstimate that a Kindle has a footprint of around 36 kg (79.2 lbs).[20] If I'm right, you'd have to go through at least thirty-six paperbacks (bought new and then sent to recycling) before the paper saving outweighs the embodied emissions of the Kindle itself. This is before accounting for electricity consumption and IT networks (for downloading the file). E-book readers may be wonderful devices, but I can't see a carbon argument for getting one, unless it gets you reading more and thus avoiding other more carbon-intense activities.

Taking a bath

200 g CO_2e modest bath heated by solar energy
500 g (1.1 lbs) CO_2e modest bath heated by efficient gas hot water tank[21]
1 kg (2.2 lbs) CO_2e generous bath heated by efficient gas hot water tank
3 kg (6.6 lbs) CO_2e generous bath heated by electricity

> A daily bath adds anywhere between 73 kg (161 lbs) and 1.1 tons CO_2e per year–that's 1.5–20 percent of the 5-ton lifestyle.

When my kids were young, at least three of our family often ended up using the same bathwater, even if not all at the same time (the muddiest had to go last). Since we topped up with hot water, the bath was always full to the overflow by the end. That was about 32 gallons, giving a footprint per person of around 330 g.

If you were to read a book in the bath for an hour, you'd probably add 50 percent to the footprint of the average full bath by pulling out the plug with your toes from time to time and topping up with hot water. So, the actual leisure activity would be 500 g per hour on top of the functional bath itself. That is quite a bit higher than watching TV, but still a lot lower than any pastime that involves using a car. In winter, you can reclaim about half the heat simply by leaving the plug in until it goes cold. This works provided you actually want the heat in your bathroom and don't mind old bathwater hanging around.

Even electric showers (see p. 50) generally out-perform baths in carbon terms. Electric showers range from 5 kW at the cheap end to 11 kW for a powerful version. For the same impact as a full bath from an efficient gas hot water tank, you could have a nine-minute power shower. Of course, it's harder to read a book in the shower.

1 to 10 kilos (2.2 to 22 pounds)

A pint (16 oz) of milk

0.42 kg CO_2e oat milk (0.9 kg CO_2e per quart)
0.45 kg CO_2e soy milk (0.97 kg CO_2e per quart)
0.9 kg CO_2e US cow's milk (2 kg CO_2e per quart)
1.5 kg (3.3 lbs) CO_2e global average cow's milk (3.2 kg CO_2e per quart)

> If your household goes through two-and-a-half pints a day, that's 850 kg (1,870 lbs) of CO_2e per year, as much as a flight from New York to Los Angeles

Cow's milk is high-carbon stuff for exactly the same reasons as beef. Cows, like most animals, waste a lot of the energy from the food they eat simply keeping warm and walking around rather than creating meat and milk. In addition, cows ruminate (chew the cud), which means they burp up methane, roughly doubling the footprint of the food they produce.[1] Finally, if the cow is fed partly on soy, there is a deforestation element to the footprint (see *Deforestation*, p. 191).

As the pie chart overleaf shows, almost 70 percent of the cow's milk footprint typically takes place on the farm, but transport, packaging, and refrigeration also play their part. Because milk is heavy, keeping it local, rather than trucking it hundreds of miles to and from distribution centers, is a good

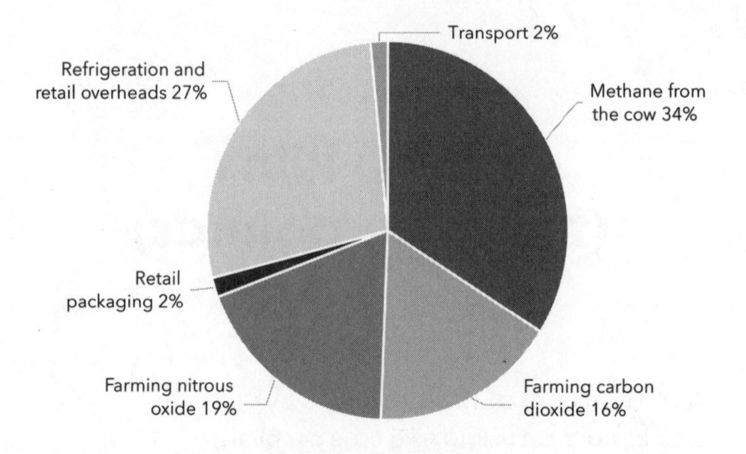

The footprint of a pint of cow's milk

idea. I've assumed that the cows are grazing on pastureland which could not be used for growing crops.

The pie chart doesn't include a slice for the journey to the store, nor for home refrigeration. In addition, reusable glass bottles almost certainly beat plastic disposables, even if you recycle the latter every time.

The estimate for global average milk is higher than for a typical US pint, due to there being more deforestation associated with it, as well as more use of cattle feed.

Wherever you get your cow's milk, though, it remains a high-carbon way to get your calories. And, if you are in a country where cows are fed on soy, it's generally a lot worse. This is a hugely complicated area to research, though. If you change the feed, you alter the carbon cost of that feed, the milk yield, *and* the amount of methane that gets belched out. Pasture-fed cows allow us to avoid the deforestation associated with animal feed but produce more methane. At the same time, you have to look at factors like the life expectancy of the cow and the amount of saleable meat the

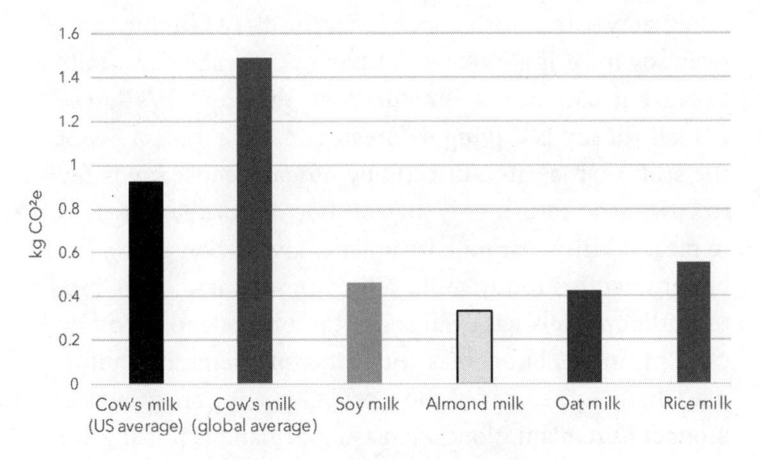

The footprint of different types of milk

herd will produce alongside its milk and inputs that will be required to keep the cow healthy. To make things even more complex, different farming practices affect the ability of the soil to absorb and store carbon. And everything also depends on the location of the farm and the breed of cow. Nobody has really worked out how all these variables interact.

If the carbon footprint were the only consideration, the most efficient thing to do would probably be to keep cattle in small indoor spaces and rear them as intensively as possible, minimizing wasteful activities such as getting exercise or keeping warm. But carbon isn't the only consideration, of course, especially for organic farmers. I interviewed a farmer called David Finlay in southwest Scotland, who reckoned that going organic reduced the milk output of his herd from 1,980 to 1,320 gallons per cow per year. He believes that, although the milk yield will go down, the amount of meat he can sell will go up and his feed costs will fall, along with his use of antibiotics and other inputs. And, of course, the animals have better lives.

Whatever the truth about different dairy farming practices, soy milk[2] is a far lower carbon option than cow's milk because it cuts out the middleman (the cow). While the market for soy is driving deforestation, the problem is not the stuff that is eaten directly by humans; most soy is fed to cows, who return only a small fraction of the nutrition in meat or dairy. Oat milk is similarly low carbon as soy, and better in coffee to my mind, although it contains less protein. Almond milk has even better carbon credentials (0.7 kg CO_2e per quart) but comes with other problems. In California, where 80 percent of the world's almonds are grown in monoculture plantations, increasing demand is putting yet more pressure on already stretched water supplies.

After a bit of getting used to it, I now prefer soy milk in oatmeal and tea.

A 250 g (8 oz) bunch of asparagus

270 g CO_2e local and seasonal

3.2 kg (7 lbs) CO_2e the same bunch airfreighted from Peru to New York

> If your entire diet were as carbon intensive as long-haul asparagus, your food would have a footprint of more than 55 tons

The numbers here are again based on data I prepared for Booths (a UK-based supermarket). They were as shocked as I was by the impact of airfreight on this kind of product and, to their credit, took steps to increase local sourcing and emphasize the benefits of local and seasonal food.

Flown in from Peru, asparagus comes in at 12.8 kg CO_2e per kilo (5.8 kg CO_2e per pound) or, to put it another way, 63 g of carbon per calorie—more than fifty times higher than bread. At the other end of the scale is asparagus grown in season in your own country. This cuts out 75 percent of the footprint, or 100 *percent if you are lucky enough to have your own patch*. And top chefs suggest asparagus is at its best eaten within forty-eight hours of picking.

None of these estimates include the footprint of cooking the food, which is likely to be around 100 g CO_2e if you simmer it for eight minutes with the lid off.

When 1 kg (2.2 lbs) of produce is moved a mile by air, it typically generates around a hundred times the carbon impact of a mile by sea. This is because it takes a lot of energy to keep a plane in the air and also because engine emissions tend to do more damage at high altitude than at ground level (see *Flying from Los Angeles to Barcelona return*, p. 152). There really is no place for airfreighted food in a low-carbon world.

Examples of other foods that are very likely, when out of season, to have been airfreighted or, just as bad, grown in an artificially heated greenhouse, include baby corn, baby carrots, edible-pod peas, green beans, fine beans, okra, shelled peas, lettuces, blueberries, raspberries, and strawberries.

When these foods are out of season, try low-carbon options such as kale, carrots, parsnips, rutabagas, or leeks. For a guide to seasonal food in your state, try the following website: seasonalfoodguide.org.

A bottle of wine

1.3 kg (2.9 lbs) CO_2e shipped from France to New York

1.4 kg (3.1 lbs) CO_2e shipped from Australia to New York, with 200 road miles

2.1 kg (4.6 lbs) CO_2e shipped from France to New York and transported 3,000 miles by road to California

> Three bottles of wine a week is equivalent to driving 350 miles in an average car

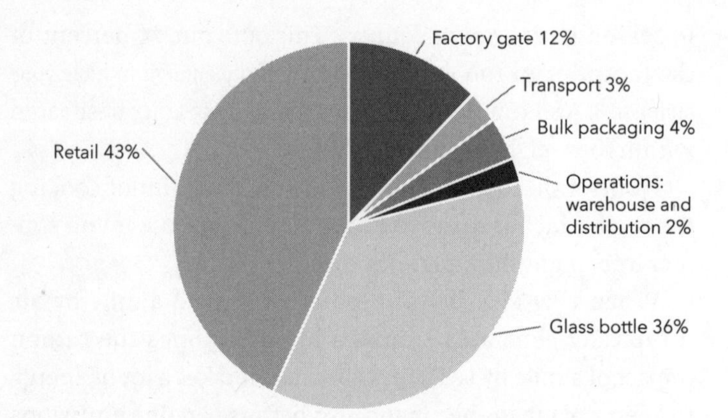

The footprint of a bottle of wine

When I set out to audit wine for Booths Supermarkets, the first thing I noticed was that the bottle typically has a bigger footprint than the wine it contains.[3] There is a simple saving to be made here if the aesthetics don't offend you: by buying wine boxes or cartons you can reduce the footprint of the packaging by a factor of about five. In doing so, you will also reduce the weight, so transport emissions can be slashed by a third. There will be absolutely no loss of quality, even though you might lose some choice. If the carton offends you, you can always decant the wine into a carafe.

But let's assume you like the choice and quality of wine in bottles. There is still a lot that can be done by the supply chain. Organico, a wine distributor near where I live in the UK, has started importing some of its wine unbottled. This cuts the transport weight. It does its own corking and puts a small deposit on the bottles, which are themselves 15 percent lighter than normal and are made from clear glass, because this is better for eventual recycling. One further nice touch is that they have done away with the concave bit under the bottle, which has always struck me as fundamentally dishonest.

Note that shipping is only a small component, so it doesn't matter that much which continent your wine comes from. Far more important are the *road miles*—a bottle of wine transported by road across the US will have the highest footprint, no matter where it was originally shipped from. For this reason, locally produced wine could cut the footprint by 20 percent, provided that your region has the right kind of climate.

Whether or not organic wine has a lower carbon footprint is unclear, although it may have other environmental benefits. And note that, because it is less dilute, wine is generally a slightly less carbon-intensive way of taking alcohol on board than beer (see A *pint (16 oz) of beer*, p. 74).

A bunch of flowers

0 CO_2e picked from your garden, no inorganic fertilizer
110 g CO_2e snapdragon, grown and sold locally
1.7 kg (3.7 lbs) CO_2e bouquet of 15 mixed stems grown outdoors and sold locally
2.4 kg (5.3 lbs) CO_2e roses grown in a hothouse, or a roses grown outdoors and airfreighted
32.3 kg (71 lbs) CO_2e bouquet of 5 hothoused roses, 5 airfreighted lilies, and 3 airfreighted baby's breath

> A bouquet a week of imported flowers could add a ton-and-a-half of CO_2e per year

The numbers above for a bouquet of imported flowers are almost the most shocking in this whole book. Out-of-season cut flowers are some of the products with the largest carbon footprint per dollar generated at the checkout. In other words, they are one of the most carbon-unfriendly ways of getting rid of your cash.

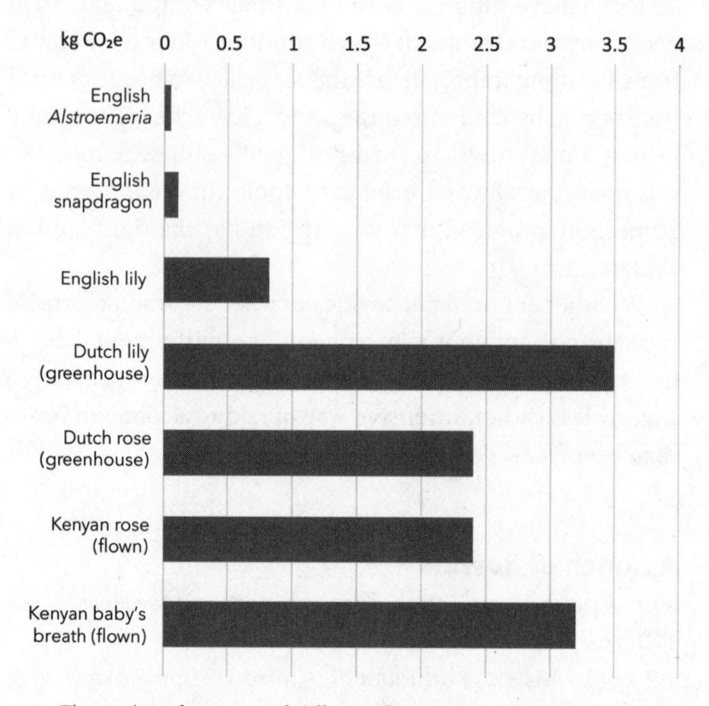

The carbon footprint of different flowers bought in the UK

They sum up the Hobson's choice (that is, an illusion of choice) that you are faced with if you want out-of-season cut flowers.[4] You either have to put them on a plane or grow them using artificial heat. Both of these are bad news for climate change. Cut flowers from your own garden are, of course, ideal. But, if that's not possible, try to go with local, seasonal flowers, which have, in the last few years, become more widely available from hundreds of small-scale grower-florists across the US, and many more across the world dedicated to the concept of "slow flowers." Small-scale mixed growing is also of huge benefit to bees and other wildlife compared to industrial monocrops.

Alternatively, opt for longer-life indoor plants, which are a dramatically less carbon-intensive option. Or go artificial—some are just about indistinguishable from the real thing.

Or perhaps try some low-carbon alternative. Could the banana ever replace the rose on Valentine's Day? Since the first edition, a good few people have got in touch with me (info@howbadarebananas.com) to say they have tried this and it can be very romantic.

There's one other concern about cut flowers, too. All commercially grown flowers use land that could otherwise be used for food. The demand for agricultural land is already driving deforestation, which in turn is responsible for between 9 and 18 percent of human-made emissions. Looked at in those terms, cut flowers have to mean less rainforest, so the true footprint is probably even bigger than my numbers suggest.

A carton of eggs

2.0 kg (4.4 lbs) CO_2e carton of 6 large eggs

> At 340 g CO_2e, an egg has the footprint of about three bananas, before you cook it

If you lived entirely off eggs (and survived the cholesterol), you'd meet your caloric needs and many times your protein requirement, for around a 3.5-ton annual carbon footprint. At 2.3 kg CO_2e per pound (or 340 g per egg), eggs are less carbon intensive than most animal products, though they are still higher carbon than most vegetable-based foods.

The pie chart overleaf shows that, as with nearly all foods, most of the impact comes from the farming rather than the packaging or transport.[5] Chickens don't ruminate, so methane isn't much of a problem. But nitrous oxide is a major

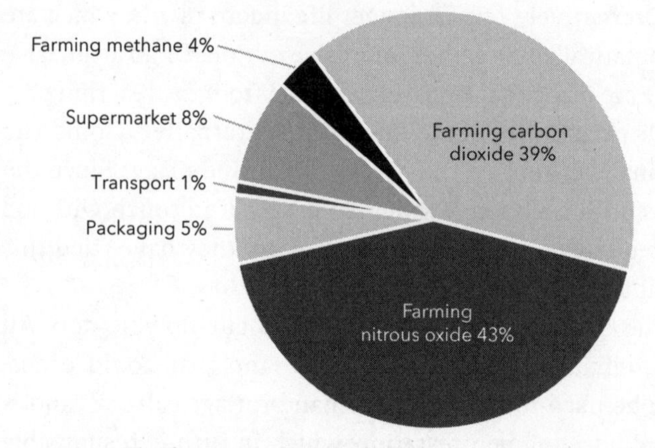

The footprint of an egg (at 340 g CO_2e each)

part of an egg's footprint, as is CO_2 from the rearing of birds and growing of their feed, which includes a contribution for deforestation. And if you've ever been inside a commercial chicken shed, you'll probably have been shocked by the heat pouring out, a visceral illustration of the inefficiency of animal farming. All that heat is energy lost in the process of turning chicken feed into eggs.

I've based my figures for the farming part of the footprint on a study by Cranfield University. This suggests that, from the perspective of climate change, organic eggs come out about 25 percent worse than those from battery farms. This goes to illustrate that, if responding to climate change sends us into a blinkered drive for efficient production, some other values may suffer.

This book isn't about telling you what values to have, but from time to time it's worth remembering that climate change is not the only issue. If you care about animal welfare as well as climate change, buying fewer eggs but making them organic might be a sensible compromise.

A day's protein (50 g/2 oz)

130 g CO_2e from 250 g of mixed nuts
220 g CO_2e from 800 g of peas
420 g CO_2e from 625 g of lentils or chickpeas
1.6 kg (3.5 lbs) CO_2e from 3 pints of milk
2.1 kg (4.6 lbs) CO_2e from 7 large eggs
2.8 kg (6.1 lbs) CO_2e from 180 g of chicken
3.8 kg (8.4 lbs) CO_2e from 200 g of pork, bacon, or ham
9 kg (20 lbs) CO_2e from 200 g of imported king prawns
10 kg (22 lbs) CO_2e from 200 g of lamb
25 kg (55 lbs) CO_2e from 190 g of beef (cattle on deforested land)

> Our diet is genuinely important

We all have slightly different protein requirements, depending on age and lifestyle, but I've based my sums on 50 g per day, which is about what the average person needs (even though in rich countries most of us eat quite a bit more than that).

The chart overleaf shows the dramatic difference in the carbon intensity of meeting our protein needs through different foods. The figures are based on global averages, although production systems vary around the world.[6]

The numbers speak for themselves, with a day's protein weighing in at less than 500 g CO_2e for all the plant-based forms—peas, nuts, soy milk, tofu, and legumes, such as lentils and chickpeas. The message for a sustainable lifestyle could hardly be clearer. The more we base our diet on plant-based food, the better. And this is getting easier and easier, even for those who like their meat sometimes, with the creation of plant-based burgers and the like from companies such as Impossible Foods and Beyond Meat.

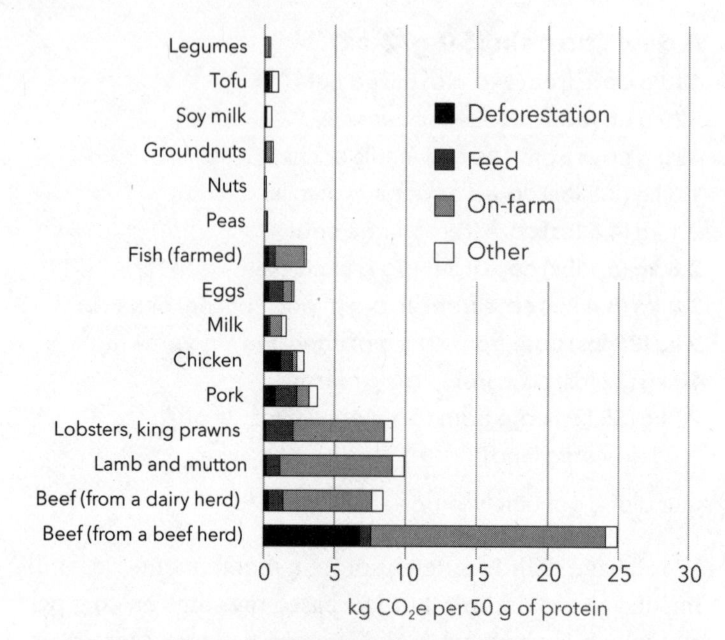

The carbon footprint of 50 g of protein

There is quite a step up in carbon terms once you add dairy products: milk (and cheese, yogurt, and cream) and eggs (slightly better). But the major impact is when you depart from vegetarian options and eat the animals rather than their products. In carbon terms, the most efficient protein comes from chickens (which grow fast, especially if chock-full of antibiotics, and don't waste much energy walking around if packed closely together) followed by other non-ruminants such as pigs. Meat from ruminants, like lamb and beef, has more than three times the footprint of chicken, and beef from cattle on deforested land is in a class of its own, especially if from a beef-only herd.

For most animal-based protein sources, the big emissions deal is what takes place on the farm, especially when the animal is a methane-burping ruminant, but also fertilizer, slurry (a mix of manure and water), and farm machinery. Where there is deforestation involved—either for land the animals will graze on or to grow the feed they eat—that is significant, as is the production of feed.

While wild fish can be pretty low carbon (despite environmental problems, see p. 78), the farmed version takes on a big footprint for more or less the same inefficiency reasons as animals farmed on land. Crustaceans—prawns, crabs, lobsters, and so on—also fall into this very high-carbon category, especially warm water prawns imported from Thailand and Vietnam, where their farming often destroys the mangrove systems, which play an important part in sequestering carbon and protecting coastlines and coral reefs.

An omission from my chart is a new kid on the food block—lab-based protein. This concept takes some getting used to, but I suspect it is on its way. If you want to create food from a given area of land, it is far more efficient to put up solar panels and use the electricity to power an entirely synthetic production system for protein and carbohydrates than it is to grow plants (or, worse, feed those plants to animals). Most plants struggle to capture more than 1 or 2 percent of the sun's energy, and only a fraction of that gets turned into food, but a solar panel can capture more than 20 percent of the sun's energy. A lab can also do a more efficient job than a plant of creating food from its energy supply.

The idea of solar panels instead of fields is pretty unappealing to me, but the upside could be the liberation of huge quantities of land for rewilding. So, we sacrifice a small proportion of the Earth's land area in order to dramatically enhance the rest. A factory in Finland already claims to be able to produce food from electricity with 20 percent efficiency.[7] That means that an incredible 4 percent of the sunlight can be turned into food energy, even though it is not yet legal to sell lab-grown meat for human consumption (except in Singapore as of December 2020).[8]

A load of laundry

390 g (0.9 lbs) CO_2e washed at 30°C (85°F), dried on the line[9]
780 g (1.7 lbs) CO_2e washed at 40°C (105°F), dried on the line
880 g (1.9 lbs) CO_2e washed at 60°C (140°F), dried on the line
2.6 kg (5.7 lbs) CO_2e washed at 40°C (105°F), tumble-dried
3.8 kg (8.3 lbs) CO_2e washed and dried in a washer-dryer

> It's the drying that creates the footprint–think about using a clothesline or drying rack, inside or out

All the figures listed above are based on a full 8 kg (18 lbs) load (half-loads use a little less energy each time but work out much less efficient per garment washed). I've added around 270 g per wash for the embodied emissions in the appliances themselves. If this estimate is correct, for a machine you keep for ten years and use efficiently, the manufacture and delivery of the appliances account for nearly 80 percent of the total carbon footprint of each wash. You can improve on the lifetime of your washer or dryer if you look after it and get it repaired when it breaks. Unlike replacing a hot water tank, switching from an old washing machine to a new one with an A rating might gain you around 10 percent in efficiency,[10] but you will struggle to ever pay back the embodied emissions. The message is that, unless your machine is particularly cranky and inefficient, keep it going.

Modern powdered detergents work just as well at 30°C (85°F), so there is a very simple saving to be had here of 400 g CO_2e per wash just by keeping the temperature down, if your machine allows it. But the bigger savings relate to drying. As the numbers above show, for a typical 40°C (105°F) wash, nearly three-quarters of the carbon footprint comes from the drying. Tumble dryers generally use electricity to generate heat. Despite the UK grid improvements, this is more than 40 percent more carbon intensive than generating heat from gas. If you use a conventional vented dryer, most of the heat is simply pumped out to the outside world, so overall it's a wasteful activity. Condensing dryers use a little bit more energy still, although (in winter at least) all that heat stays inside your house, where it is probably of some use.

Overall, a household running the tumble dryer 200 times a year could save 400 kg (880 lbs) CO_2e by installing a clothes rack inside and a clothesline outside. In winter, the evaporation from drying clothes will cool your house down slightly, but it's a marginal effect and on a baking hot summer's day our clothes drying in the kitchen act as free air conditioning. If you must use a dryer, make sure your washer has a good spin function. It is much more efficient to remove most of the water by spinning it off than by evaporating it.

While on the subject of washing, have you optimized the frequency with which you wash stuff? I don't want to promote smelly clothes, but it's worth asking yourself this question: does your stuff go in the wash unnecessarily often? If you can reduce the number of washes without upsetting anyone, there is a time saving to be had, too, so it's a great example of life getting better as the carbon comes down.

Desalinating 1,000 liters (264 gallons) of water

400 g CO_2e solar-powered reverse osmosis[11] plant, brackish water

2.3 kg (5 lbs) CO_2e reverse osmosis plant, conventional media filtration, seawater

6.7 kg (14.7 lbs) CO_2e reverse osmosis plant using electricity from coal, seawater

34.7 kg (76 lbs) CO_2e inefficient thermal desalination plant

> Desalination may account for 0.4 percent of the world's carbon footprint

Every day the world desalinates around 25 billion gallons of water—that's 95 million tons or cubic meters, up from 16 billion gallons in 2010 and rising by 10–15 percent a year.[12] Something like half of the total takes place in and around the dry, oil-rich Middle East, but desalination also accounts for a large chunk of electricity usage and gas consumption in California. Emissions per gallon vary hugely depending on the efficiency of the process and the carbon intensity of the electricity used.

The global carbon footprint is between 80 and 230 million tons CO_2e every year, or something like 0.15 to 0.4 percent of all global emissions.[13] This figure is likely to continue rising, not least because the world is getting hotter and drier in many regions as a feedback loop of our climate emergency.

At the high end of the spectrum are inefficient thermal desalination processes. A big improvement on this is reverse osmosis. Various options exist for using spare heat from fossil-fueled power stations (though we should be careful not to double-count the benefits: in these situations, the desalination plant may claim carbon neutrality, while the power station claims to be offsetting its emissions by supporting the desalination plant).

At the low-carbon end is reverse osmosis desalination of brackish water, which contains less salt. Note that, as the biggest component is the electricity use, the overall carbon footprint is highly dependent on the source of electricity. So, a reverse osmosis system powered by solar energy is at the lower end of the spectrum.

Seawater Greenhouse claims to have developed a technique for using solar heat to desalinate water for greenhouse-cultivated crops in arid regions.[14] In theory, the desalination itself is just about carbon neutral. I haven't personally investigated the technique, but the company has won awards and has some large pilot projects already in operation. It is the kind of technology that gives hope in the midst of increasing desertification problems around the world.

Looking ahead, emerging technologies using nanoporous carbon composite membranes, such as graphene, might stand to lower energy consumption by as much as 80 percent compared to conventional membranes.[15]

Apart from greenhouse gases, another nasty and common byproduct of desalination is the brine concentrate that is returned to the sea, increasing the salinity and messing up marine ecosystems.

250 g (8 oz) of cheese

1.6 kg (3.5 lbs) CO_2e goat's cheese
3 kg (6.6 lbs) CO_2e cheddar
4.8 kg (10.6 lbs) CO_2e parmesan

> Dairy is high carbon–a block of cheddar is equivalent to a massive 10 kg (22 lbs) of carrots

It takes about 10 quarts of milk to make 1 kg (2.2 lbs) of hard cheese, adding up to a carbon footprint that's higher than that of many meats. The message is clear, then: going veggie doesn't reduce your impact if you simply eat cheese instead of meat. Neither will it save you money or make you healthier. Perhaps the best advice if you're eager to reduce the climate impact of your diet is to think of cheese as a meat and therefore a treat. Many people will also improve their life expectancy by cutting back somewhat.

However much cheese you eat, there's an easy carbon win by keeping waste to a minimum. That means buying only what you think you'll actually get through and not throwing

Type of cheese	Yield (percent)	Quarts of milk required for 1 kg of cheese	Total footprint (kg CO_2e/kg)
Cottage	20	5	6.50
Brie	14	7	8.80
Mozzarella	11	9	10.8
Cheddar	10	10	11.8
Parmesan	6	17	19.1

The footprint of some popular cheeses

it out if it's showing a tiny sign of mold. You just need to trim off the mold on hard cheese. Soft cheese is more complex: some rind molds are integral to a cheese and entirely benign; others may be more suspect. You'll extend its life if you cover cheese in cling wrap and keep it refrigerated.[16]

As for which hard cheese to buy, the most sustainable types probably come from sheep, goats, or cows that have grazed almost exclusively on rough pasture that couldn't have been used for crops, although of course that information isn't generally available in the supermarket or at the deli counter.

Note that which country or area the cheese has come from doesn't matter much when set against the impact of the milk production (see A *pint (16 oz) of milk*, p. 83). Hence, the easiest way to reduce the carbon footprint of your cheese is to opt for soft cheeses, because they require less milk to produce. Goat's milk only requires 5 quarts of milk to produce 1 kg (2.2 lbs), whereas parmesan requires 17 kg (35 lbs). The table above shows the amount of milk required to make 1 kg (2.2 lbs) of cheese and the overall carbon footprint.

Finally, vegan cheeses (even though these are not allowed to go by that name) are getting better all the time. And, if you're vegetarian, you probably ought to know that cheese production (even with vegetarian rennet) still involves the slaughter of animals. Dairy animals need to be kept in a milk-producing state. And farmers don't put the male calves or kids or rams out to pasture.

A 125 g (4 oz) burger

360 g CO_2e vegan bean burger (no cheese)
630 g CO_2e veggie bean burger (with cheese)
6.6 kg (14.5 lbs) CO_2e cheeseburger

> If you ate a 4-ounce cheeseburger each day, that'd be 1.2 tons CO_2e per year—three months' worth of a 5-ton lifestyle

A 4-ounce cheeseburger provides 515 calories. If this were the only type of food you ate, you would need about five burgers per day, provided you didn't do much exercise.[17] If you managed to keep up this diet for a year (without killing yourself), you'd cause about 6 tons of carbon emissions.

As the chart overleaf shows, both the beef, and, to a lesser extent, the cheese, are the key factors. As we've already seen, animal produce tends to be more carbon intensive than vegetables and grains because animals consume a lot of energy, which makes their conversion of animal feed into meat and milk inherently inefficient.

There is another big problem with beef and dairy farming. Cows, like sheep (but not pigs), are ruminants. This means they belch out methane, a greenhouse gas twenty-eight times more potent than CO_2. The result is that beef and lamb have around double the carbon footprint per kilo of meat compared with pork (see *A day's protein (50 g/2 oz)*, p. 93). A further consideration is that excessive demand for meat provides an incentive for deforestation (see p. 191) because

Component	Grams CO_2e
Beef (108 g)	2,840
Cheese (20 g)	270
Bun (40 g)	47
Salad (20 g)	28
Condiments (20 g)	50
TOTAL	3,235 (+ extra for cooking)

The footprint of a 4-ounce cheeseburger

it raises the demand for arable and grazing lands. That said, there is plenty of land, for example in the UK, that is fit for cattle and sheep farming but not for crops, and there can be a conservation benefit in having animals on the land.

It is unclear whether the footprint of the burger could be reduced by using organic or free-range meat and cheese. A Cranfield University study[18] found that organic cattle farming had little or sometimes negative carbon benefits. However, the organic farmers I know who have studied this report made scathing comments on the assumptions made about organic practices and yields. The carbon benefits of rough grazing are also unclear: less feed is required, but there are complex implications on yield and rumination.

However you look at it, there is not much room for beef in a low-carbon lifestyle. Luckily, the veggie and vegan alternatives are now superb, you can have five to ten times as many burgers for the same footprint, and once you start to eat them you'll get to prefer them.

1 kg (2.2 lbs) of rice

3 kg (6.6 lbs) CO_2e efficiently produced

4 kg (8.8 lbs) CO_2e average production

7.1 kg (15.6 lbs) CO_2e poor production with excessive fertilizer

> A kilo of rice can cause more emissions than burning a quart of diesel

Rice deserves a place in your consciousness not only as a food on your table but as an important piece of the global food and carbon jigsaw. Europeans and Americans get just 1 or 2 percent of their food energy from this crop, but the proportion is very much higher in Asia, where 89 percent of the world's total rice harvest is consumed.[19] Globally, it provides 20 percent of the world's food energy in exchange for 3.5 percent[20] of its carbon footprint.

I suspect that many of us will be surprised and unsettled to hear that rice, the simplest of foods, is a surprisingly high-carbon staple, much more so than wheat, which is nutritionally similar. That's because of the methane that bubbles out of the flooded paddy fields and the excessive nitrogen fertilizers that are all too often applied.

Around the world, 600 million tons CO_2e of methane is thought to be emitted from rice paddies, accounting for around 1.2 percent of the total global carbon footprint (for comparison, this is about three times the footprint of all the cement produced in Europe). Even more significant are the 167 million tons of fertilizer, mainly nitrogen-based, that are applied to the crop. That's a little over 1 ton of fertilizer for every 3 tons of rice produced. If fertilizer is made in an efficient factory and applied sparingly, at well-chosen moments, each ton may only result in 2.7 tons CO_2e. If not, the figure could be as high as 12.3 tons CO_2e.

I have guesstimated 300 g CO_2e per kilo of rice for the production of agricultural machinery and the transport of

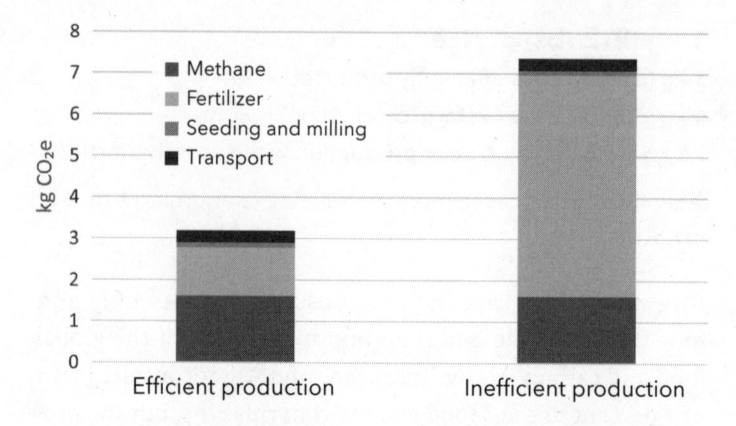

The carbon footprint of 1 kg (2.2 lbs) of rice

the rice to market. Most is eaten in its country of origin, and I can't imagine rice ever finding its way into a hothouse or an airplane.

It's possible to grow rice without flooding the field and thereby cut out the methane. However, it's harder work (you have to do more weeding) and may require more fertilizer, which could end up tipping the net carbon impact the wrong way. As with much agriculture, we don't fully understand what goes on or what the best options are; this is another important area for research, as the number of human mouths soars, along with the global temperature.

If we were to feed the world entirely on food as carbon intensive per calorie as rice, emissions from food farming would be halved (assuming rice grown in the most efficient manner). But if the worst rice-growing practices were the norm, farm emissions would increase.[21]

A takeout taco

4 kg (8.8 lbs) CO_2e vegan bean and avocado tacos for 4
10 kg (22 lbs) CO_2e veggie bean and cheese tacos for 4
33 kg (73 lbs) CO_2e beef and cheese tacos for 4
+ 340 g CO_2e per mile for a delivery by scooter, more by car
+ 13 g CO_2e for each aluminum tray with lid (10 times higher if you don't wash and recycle it)
+ 25 g CO_2e for each clear plastic tray

> Tacos have a hugely diverse footprint, depending on your order

A takeout taco comes in at anywhere between 1 and 8 kg (2.2 and 17.6 lbs) CO_2e per person, depending on whether your order is primarily vegetarian and whether you buy too much. A vegan bean and avocado taco can be as carbon efficient as any cuisine. If you have cheese instead of avocado, that adds an extra 1.5 kg (3.3 lbs) CO_2e per person, more than doubling the footprint. But the real carbon dealbreaker is if you swap out the beans for beef. This adds a further 6 kg (13.2 lbs) CO_2e and makes your takeout more than eight times worse than the vegan version.

The most important message is to go veggie as much and as often as possible, which has the added advantage of avoiding meats whose origins you probably know little about. Next most important is to avoid the huge temptation to order twice as much food as anyone can eat. Transport is not a big part of the footprint but, if possible, walk down to the taco stand (and in doing so work up an appetite), rather than getting them to deliver. Finally, don't be embarrassed to bring your own reusable containers.

For the taco dishes, I've used a couple of recipes from the web; one to make the crispy taco shell and one to make Taco Bell–style tacos.

1 kg (2.2 lbs) of plastic

1.7 kg (3.7 lbs) CO_2e PET plastic bottles, made from recycled materials and recycled again after being used[22]

3.4 kg (7.5 lbs) CO_2e polystyrene from virgin materials

4.2 kg (9.2 lbs) CO_2e PET plastic bottles, made from virgin material, sent to landfill after being used

5.4 kg (11.9 lbs) CO_2e PET plastic bottles, made from virgin material, incinerated after being used

9.1 kg (20 lbs) CO_2e some types of nylon[23]

> The world generates 1.8 billion tons CO_2e from plastic production every year—that's 3 percent of annual emissions, and counting...

Plastic is such useful stuff: it's tough, durable, and waterproof. No wonder we use so much of it. Unfortunately, these same qualities mean that it hangs around in landfill sites for centuries, clutters up the stomachs of animals and fish, transforms remote beaches into junkyards, and has ended up in almost every ecosystem you can think of. We are also only just starting to realize the effects this might be having on our own health, as well as that of other living things.[24]

From a purely carbon perspective, plastic's inability to rot is good news in that it won't add to methane emissions from landfill. If we assume that the plastic is put in the trash rather than tossed into a street or field, those hydrocarbons are going back underground where they came from. However, by throwing plastic into a landfill instead of recycling, we are forcing yet more virgin production (see *1 kg (2.2 lbs) of trash to landfill*, p. 71). The chart opposite shows the emissions from different kinds of plastic made from virgin and recycled materials.

Despite increased awareness of the problems of plastic, not least through David Attenborough's efforts, global

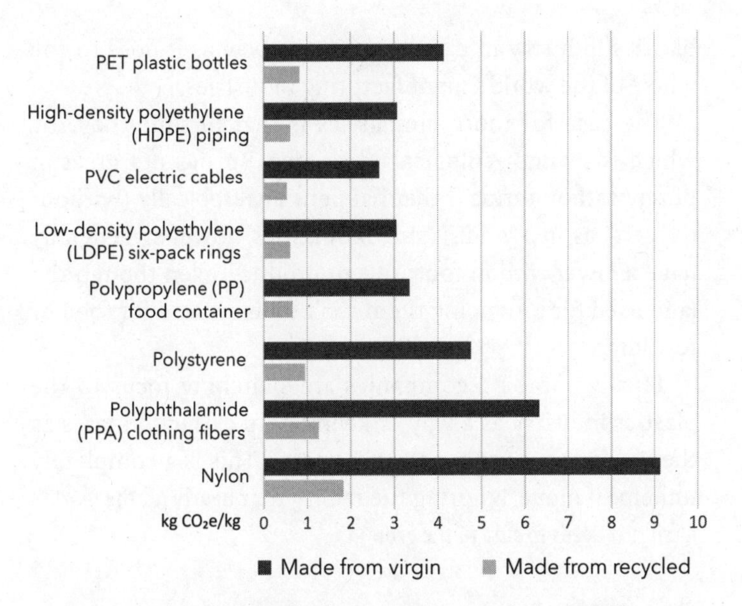

Types of plastic made from virgin or recycled material

production is still increasing by a shocking 4 percent every year. Cutting demand is tricky because it is so convenient that it has found its way into every corner of our lives. But there is no getting around the need to try to untangle ourselves from it wherever we can.

There are four things we need to do:

- **Increase recycling rates massively** from a pitiful 18 percent.[25]
- **Increase the proportion of renewable energy** used in production.
- **Increase the proportion of bioplastics** in the mix.
- Most importantly, **use less plastic**.

Plastics recycling is tricky, as different types have to be separately processed. This is a particular challenge when products blend different types of plastic together, as many do.

Increasing the renewable energy in the mix applies to the

plastics industry in exactly the same way as it does to the whole of the world's manufacturing industries.

The case for more bioplastics is that in time they rot, which is a huge ecological advantage. But it's not great in purely carbon terms if that happens anaerobically (without oxygen), as in a landfill site. Bioplastics, however, generally have a lower carbon footprint to produce, even though the land used for producing them can't then be used for food or rewilding.

Finally, some oil companies are switching focus to the plastics industry as a way of keeping up extraction rates as the world moves to renewable energy. This is a completely unhelpful move, ignoring the reality that *nearly all the world's fossil fuel needs to stay in the ground.*

1 kg (2.2 lbs) of tomatoes

1.3 kg (2.9 lbs) CO_2e large salad variety, grown locally, in season[26]

4.9 kg (10.8 lbs) CO_2e baby plum, grown outdoors in summer

54 kg (119 lbs) CO_2e organic vine cherry tomatoes, grown in a heated greenhouse in Ohio in March

> Beware hothouses

That's a shocking headline: tomatoes, when hothoused, are one of the highest-carbon foods in this book. But there's good news, too. Summer tomatoes are fine. At the low end of the scale, a high-yield classic variety is grown in the summer with no artificial heat required. These loose tomatoes, the ones that our parents were brought up on, cause only just over half the carbon of specialist varieties (defined here as cherry, plum, cocktail, beefsteak, and others), because the yield is so much higher.

Sadly, the smaller and tastier varieties have lower yields

and turn out much higher carbon per kilo, although perhaps not per unit of taste. At the high-carbon end, out-of-season organic cherry tomatoes sold "on the vine" can be responsible for a staggering 54 kg of greenhouse gas per kilo.

These high figures come from a detailed but controversial study by Cranfield University.[27] Perhaps their most unsettling finding was that, when heat from fossil fuels is required, organic is the highest-carbon option, again because the yield was thought to be lower.

So, tomato lovers concerned about climate change would do well to stick to the season and to favor classic varieties, sold loose. In the winter, it makes carbon sense to stick to canned tomatoes, but if you want to buy fresh tomatoes outside the local growing season it is almost certainly preferable to buy them from California, Mexico, or another warm place that isn't too far away, rather than local versions produced in heated greenhouses.

A 250 g (8 oz) steak

12.1 kg (26.6 lbs) CO_2e raw steak, from the US
17.8 kg (39.2 lbs) CO_2e raw steak, from deforested land in Brazil

> Eating a US-reared steak has about the same impact as 100 bananas; if it comes from Brazil, it may be more like 150 bananas

Beef is a climate-unfriendly food, weighing in at nearly 53 kg CO_2e per kilo (24 kg CO_2e per pound) in the US.[28] The global average figure is higher still, as a lot of cattle are reared on deforested land and fed fodder, which is also commonly produced on deforested land.

For a US steak, 98 percent of the footprint comes from the beef farming itself. As we've already seen, using animals to produce food tends to be inefficient compared with eating crops, and cows have the added problem that they ruminate, producing enough methane to roughly double the climate change impact of farming them.

Less widely discussed than the methane are nitrous oxide emissions, which typically account for about three-tenths of the footprint of beef farming. This gas is released when nitrogen fertilizer is applied to grass and other fodder crops, and when the grass is silaged. Finally, there is the CO_2 itself, at around one-fifth of the farming footprint. This is caused by the tractors, other farming machinery, and energy required to make fertilizer.

I'm using the same footprint figure here for all beef, although you could argue the case for attributing more to the most expensive cuts than to the mechanically reclaimed stuff that finds its way into economy burgers. In that sense, offal and processed meat may well be a greener choice than more premium meat products. But however you look at it, food from cows is at the top end of the carbon spectrum, despite ongoing progress to mitigate emissions (see *A pint (16 oz) of milk*, p. 83, and *A 125 g (4 oz) burger*, p. 101).

If you cook your steak on an outdoor disposable barbecue, you've added to the problem with a particularly unsustainable form of cooking: high emissions from burning the charcoal, and aluminum that is very hard to recycle after it has been covered in burned animal fat. One option, if you've already got one of these lying around, would be to take it apart, crunch up the charcoal and bury it as biochar (see p. 204), then recycle the clean aluminum.

10 to 100 kilos (22 to 220 pounds)

A pair of shoes

1.5 kg (3.3 lbs) CO_2e rubber Crocs
8 kg (17.6 lbs) CO_2e synthetic
11.5 kg (25.3 lbs) CO_2e average mix
15 kg (33 lbs) CO_2e all leather

> Imelda Marcos's collection of 2,700 pairs of shoes had a footprint of around 30 tons[1]

Shoes vary enormously in their carbon footprint (no pun intended), depending on what they're made from and how long they last. At the lowest end of the carbon scale are Crocs, the simple and surprisingly durable shoe consisting of just 250 g of expanded EVA (foam rubber) and sold without packaging. For these shoes, the raw material comes in at just over 1 kg (2.2 lbs) CO_2e.

The 8 kg (17.6 lbs) CO_2e synthetic pair is based on a study of synthetic fell (off trail) running shoes, made in China but traveling to the UK by boat.

I used an input-output model to estimate the footprint of all leather shoes and assumed that in the typical shoe about half the carbon footprint comes from the materials, a quarter from energy used in shoe manufacture, 15 percent transport, 5 percent the shoebox, and another 5 percent of bits and bobs.[2]

Most of our footwear comes from East Asia, although specialist leather might also have had to travel a long way to get there.

Shipping is fairly efficient. The big inefficiency in transport comes if a product is airfreighted for speed, which is most likely in high-end fashion. Unfortunately, products don't carry labels with freight information, so there's no way to be sure of their status. This ought to change.

A pair of pants

8 kg (17.6 lbs) CO_2e polyester pants (300 g)
11 kg (24.2 lbs) CO_2e acrylic pants (300 g)
19 kg (41.9 lbs) CO_2e men's cotton pants (600 g)

> "Natural" materials may sound greener, but the footprint tells a different story

My cotton pants weigh 600 g. Recent studies have found that their production has a footprint of around 26 kg CO_2e per kilo (11.8 kg per pound).[3] That includes the dyeing, cutting, and sewing, an allowance for waste fabric, buckles, zippers, and so on, as well as transport, meaning that by the time you buy them from a store a single pair of pants will have a footprint of around 19 kg (41.9 lbs).

But this figure doesn't tell the whole story. Their lifetime of wearing and washing adds 12 kg (26.4 lbs) CO_2e per kg (this assumes a tumble dry after each wash). Add in another 1 kg (2.2 lbs) CO_2e for any pants that end up in a landfill at the end of their lifetime and the footprint goes up to 32 kg (70.4 lbs) CO_2e per pair of pants. You can bring down this figure by washing them less and drying them on the line, but the footprint will still add up to over 20 kg (44 lbs) CO_2e.

For synthetic fibers, emissions for the production phase

are even higher per kilo than for the cotton pants. But, crucially, they are half the weight and so the footprint per pair is lower. The use phase is lower too, as the synthetics take up less of the washing machine and will easily dry on the line. If well made, they will also last significantly longer than the cotton pants.[4]

Even if I wear my cotton pants right into the ground, I can't envisage getting more than 200 days of solid use out of them. That works out at a minimum of 95 g CO_2e per day, or more than 160 g per day once I factor in the laundry and disposal.

By comparison, the synthetic pants are probably good for 600 days of wear, and because they virtually drip-dry the laundry aspect will likely only be a quarter that of the cotton pants. So, all told, that's just 23 g per wear-day for the polyester and 28 g for the acrylic (see A *load of laundry*, p. 96). This means the synthetic pants are between six and seven times less carbon intensive than the cotton pants.

Globally, around 100 million tons of textiles are produced per year, almost double the total in the early 2000s and increasing at a rate of 2 percent per year.[5] Once you take into account the full life-cycle emissions, that equates to 3.3 billion tons CO_2e per year, or around 6 percent of global emissions.[6] The fashion industry also consumes 21 trillion gallons of water per year. The Aral Sea has dried up in part because of cotton plantations in its catchment, and the clothing and textiles industry produces toxins that find their way into water supplies.

I estimate that if you live in the US, clothing and textiles will typically make up about 2 percent of your footprint. But if you buy a lot of clothes—fast fashion that only gets worn a handful of times—your footprint may be five or ten times that figure.

Here are some tips for keeping the total impact of your clothing to a minimum:

- Buy stuff that is easy to wash and dry.
- Buy stuff that is built to last, wear it, repair it, and use it until it falls apart (or pass it on).
- Buy secondhand.
- Donate or recycle clothing rather than putting it in the trash.
- Favor synthetic fibers over natural ones.

A rush-hour car commute

16 kg (35.2 lbs) CO_2e 10 miles of crawling to and from work by car

> A stop-and-start drive in gridlock traffic can cause three times the emissions as the same drive on a clear road[7]

Driving on a congested road very roughly doubles your own fuel consumption per mile. However, that's only half of the story. By adding your car to the mass of belching motors, you also make a lot of other people take just a little bit longer in *their* cars. It turns out, via a bit of simple queuing theory,[8] that the extra emissions you force everyone else to produce are about equal to the extra emissions that you produce yourself. In other words, if your journey is badly congested, you cause about *three times more emissions* than if you had driven on an open road.

The queuing theory logic also works for the time wasted. If you make the assumption that the journey is many times longer than it would be if there were no traffic, then the time you waste in the queue is about equal to the sum of the extra time you make everyone else waste. In other words, the hassle and anguish that you experience is equal to the hassle and anguish that you inflict.

All of this adds to the case for traveling by bike, bus, train, foot, or ride-share wherever possible. It's also a useful reminder that motorists should treat cyclists with the respect they deserve for helping to cut everybody else's emissions and wasted time.

If you have to drive in busy conditions, do your best to minimize stops and starts—both your own and everyone else's. Emissions-wise, a steady, slow stream of traffic is more efficient than a stop-and-start traffic jam, unless the stops are so long that everyone can turn their engines off. One tip is to think about what to do when two lanes merge: to avoid anyone having to queue, ease your speed down, merge gently and in good time, and allow others to do likewise. In theory at least, two lanes traveling at 50 mph can carry about the same traffic as three lanes traveling at 70 mph, assuming everyone leaves a safe stopping distance between them and the next vehicle. This is because slower cars need less distance between them.[9]

It's good to minimize the use of brakes on the highway if you can. And, when you overtake, put your indicator on in good time so no one else has to brake.

Finally, COVID-19 has shown many of us that working from home is often more possible than we thought. If we can keep the habit, it has to be good for carbon, air quality, and for cutting the amount of time we all waste in traffic jams.

A bag of cement (25 kg/55 lbs)

9 kg (19.8 lbs) CO_2e one "green" cement alternative
17 kg (37.4 lbs) CO_2e industrially trialed low-clinker cement
24 kg (52.8 lbs) CO_2e global average ordinary Portland cement

> Cement production makes up 4.5 percent of the world's greenhouse gas footprint

The world produces just over 4 billion tons of cement per year—a staggering 525 kg (1,155 lbs) for every person on the planet. And this number is rising fast[10] with the rapid urbanization of nations such as India and Indonesia. The emissions are so high because the chemical process that turns limestone into cement gives off large volumes of CO_2 and also takes a huge amount of energy to produce. The figures above are for a standard 25 kg (55 lb) bag of cement that you could pick up at your local builder's yard.

Around half of cement's footprint comes down to the chemical reaction. There is not much you can do to reduce this without changing the product itself or reducing the amount of clinker (the main ingredient of cement that gives off the emissions). About 40 percent comes from the burning of fuel to drive the reaction. The other 10 percent is other elements of the cement industry and its supply chains.

Cement makes up about 12 percent of the footprint of the UK construction industry (and a similar proportion in the US), so other potential ways of reducing its impact are to use different materials, and to build to last and to refurbish, in preference to knocking down and building anew (see A *new-build house*, p. 167).

There are no obvious silver bullets for reducing emissions from cement production, but somehow the construction industry is going to have to find ways of doing so over the coming years. One way of reducing the footprint of cement is to decrease the amount of clinker by adding supplements like volcanic ash (which is what the Romans used), fly ash, and ground granulated blast slag. An industrial trial using these tactics has been carried out in Cuba, which reckoned a 30 percent reduction in CO_2e compared to global average cement.[11]

Another alternative is to use novel cements with alternative starting materials. According to one study, alkali-activated metakaolin- and calcium hydroxide-based cements have a 50 percent lower footprint (360–380 g CO_2e/kg), as well as a higher thermal conductivity and better recyclability.[12] Magnesia oxide-based cements may be even better. These cements absorb CO_2 during the hardening process, theoretically enabling them to be carbon negative for the process emissions. Despite their potential, however, these novel cements have not yet reached widespread application.

Leaving the lights on

28 kg (62 lbs) CO_2e 5-watt low-energy bulb for one year
570 kg (1,254 lbs) CO_2e 100-watt incandescent bulb for one year

> LEDs have made a huge difference–but make sure you've switched over

Leaving a light on for a whole year might sound extreme but having an average of one bulb turned on unnecessarily at any one time is almost certainly quite common.

Low-energy bulbs have the potential to save an enormous amount of electricity. By incandescent bulb, I mean the old-fashioned kind with the glowing tungsten wire. In the UK, it is now illegal for stores to buy them, so they are becoming museum pieces—at last.

However, efficiency alone won't bring about a low-carbon world because the less costly something becomes, the more we tend to use it, so the result can end up being more consumption. In the case of lighting, this translates, for example, into having twenty lights in the kitchen and thinking "I've left a few lights on but it's okay because they are low-energy ones." There's also the fact that the money we spend on bills will end up being spent elsewhere—a cheap flight, perhaps (see discussion of the rebound effect, p. 18).

Like any form of electricity wastage, the precise impact depends on where you live. I've based the figures here on a typical US energy mix. You could argue that in France it's okay to leave the lights on because it mostly comes from low-carbon nuclear power, but in my analysis that doesn't stack up (see *A unit of electricity*, p. 55).

Finally, there's no truth in stories you may have heard that the act of turning a light on uses the same energy as leaving it on for half an hour.

A night in a hotel

3 kg (6.6 lbs) CO_2e bed-and-breakfast in a nice hotel that takes its footprint seriously

30 kg (66 lbs) CO_2e dinner, drinks, bed, and breakfast in a hotel with average eco-credentials

75 kg (165 lbs) CO_2e night in a lavish hotel with no eco-credentials

> Carbon footprints follow you on holiday...

Let's look first at the high-carbon scenario. It's one of those hotels where the TV and six lights are already on when you walk into your room. The room itself is too hot and you cool it by opening the window even though the radiator is on. There is a swimming pool with air conditioning. You order beef or lamb for dinner and it arrives with baby vegetables airfreighted from Peru. There is too much for you to eat. For dessert, you have strawberries, even though it is winter. In the kitchens, half of the food cooked is thrown out at the end of the night. You stay one night, finding your way through three towels and a dressing gown, as well as your sheets. You have a full breakfast in the morning, giving the paper you ordered a quick glance before leaving it on the table (from

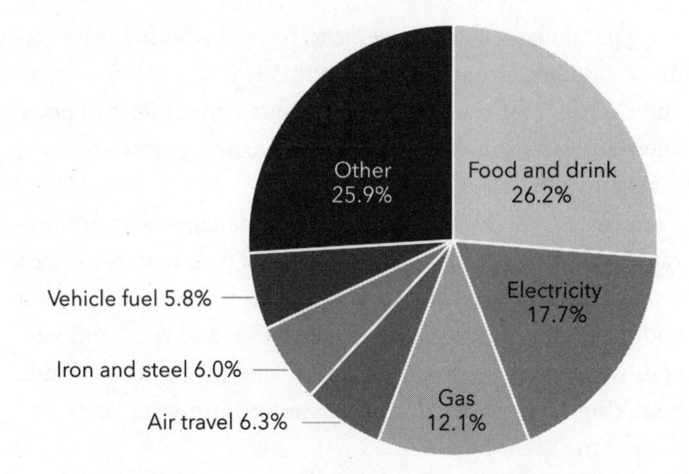

Breakdown of accommodation sector's footprint in 2016

where, surely, even in this hotel, it goes for recycling). It's wrong on all counts.

At the low end of the scale could be a large, very well-run hotel or, more likely, a simple bed-and-breakfast with thoughtful owners. If you stay a few nights, your sheets and towel aren't changed unless you ask. The room is comfortable, and you can adjust your own heating. You opt for a meal low on meat and dairy, with seasonal vegetables, and the portion isn't excessive. Kitchen leftovers end up in the next day's soup. You have something like cereal or muesli with a plant-based milk, fresh fruit, and toast for breakfast. There is a selection of papers shared between guests.

The difference in carbon footprint between these two scenarios might be as high as a factor of twenty-five.

The British clock up about 9 million tons of emissions through our use of hotels, pubs, cafés, and restaurants (see pie chart above). That's more than 1 percent of the national carbon footprint. As a rule of thumb, the carbon footprint is about 300 g CO_2e for every pound we spend (or roughly

222 g per dollar), and this seems to be true whether is it food, drink, or a hotel room. However, this is just a general figure and there's a lot the carbon-conscious consumer can do to keep emissions down, simply by favoring establishments that think about the issues.

Look for a hotel with good energy management, minimization of laundry, and a general sense of care with resources. Its restaurant, for example, should offer low- or zero-meat and dairy options, cooked with passion, and fruit and vegetables that are seasonal or from frozen. The hotel should be taking steps to minimize waste both on your plate and behind the scenes.

A leg of lamb (2 kg/4.4 lbs)

65 kg (136.4 lbs) CO_2e joint at the supermarket checkout (global average)

> For the same carbon footprint, you could have a bowl of oatmeal every day for four-and-a-half months

Globally, lamb comes with a carbon footprint of about 30 kg CO_2e for each kilo (13.6 kg per pound) produced at the slaughterhouse. Transport, basic processing, refrigeration, and a little bit of packaging each add a little extra, so that by the time the meat reaches the checkout the footprint has increased by about 10 percent, taking the overall carbon impact to more than 32 kg per kilo (14.8 kg per pound).

The issues surrounding sheep are very similar to those relating to cows (see A 250 g (8 oz) *steak*, p. 109, and A *pint* (16 oz) *of milk*, p. 83), as sheep are also ruminants, burping up large quantities of methane. However, as with beef farming, the impact of different sheep farming practices is complex and only partly understood. Hill farmers can claim that they are putting otherwise unproductive land to use and that animal quality of life is high. Counterarguments include that hill-farmed sheep are inefficient because they spend so much energy wandering around, eating low-energy food, and keeping warm and thus burp more methane per joint of meat than lowland sheep. It is also clear that overgrazing strips out much of the biodiversity and, since sheep on overgrazed land ruthlessly nibble any new shoots as soon as they pop up, even a fairly low stock rate is enough to prevent recovery.

An analysis I carried out for one farm in the UK concluded that there was more inherently human-digestible nutrition in the sheep feed than in the animals that the farm sold (although I do take the point that most people wouldn't want to sit down to a Sunday roast of sheep feed). From a financial point of view, we found that the cost of feed and medicines also turned out to be more than the sale price of the sheep, meaning that the farm survived only through subsidies, and everyone would have been better off if the farmers had been paid the same amount *not* to keep sheep.

The Lake District recently won World Heritage status, partly on the back of its sheep farming traditions, but on the farm I looked at it turned out that the actual practices aren't quite what many tradition-lovers expect. As well as eating imported feed and medicines, these sheep went on a truck to the south of England to spend the second half of their lives fattening up for slaughter. I'm not against all sheep farming, and I definitely want to see better support for all the UK's farmers and their communities. However, this needs to be done in a way that is respectful of the messages from science. The farmers I've talked to get this and just want the right support and advice.

It is great to say that we need to subsidize the goods and not the bads but delivering on that will mean getting right into the detail, with high-quality collaborative discussions. Every patch of land on a different farm is a bit different and needs detailed analysis of the options: whether to grow trees, let it go wild, plant crops, or raise animals—and exactly how best to go about it.

A week's food shopping

17 kg (37.4 lbs) CO_2e vegan, no airfreight, no waste
42 kg (92.4 lbs) CO_2e vegan, average
53 kg (116 lbs) CO_2e vegetarian, average
81 kg (176 lbs) CO_2e average diet, including meat
121 kg (266 lbs) CO_2e average diet, including meat, airfreight

> If you have a meat-heavy diet in the US, you could use up 125 percent of a 5-ton lifestyle just on food and drink

There is no getting around it: the more meat and dairy you eat, the higher your carbon footprint will be. A vegetarian diet in the US saves 34 percent of carbon footprint and a vegan diet can save nearly 50 percent. Equally important is avoiding foods that have been airfreighted in. Just avoiding these can knock anything between 13 and 40 percent off your diet.

Vegans and vegetarians, who naturally eat more fruit and veg, should be especially wary of airfreighted foods, as these are the ones most likely to go by air. Green beans, asparagus, peas, salad vegetables, and fruits like berries, mangoes, and pineapples are among the most common. I haven't factored hothousing into the equation, but this is nearly as bad as flying in terms of emissions. The easiest way to avoid foods that have been hothoused or airfreighted is to go seasonal as much as possible (see seasonalfoodguide.org for seasonal food in your state) and for stuff out of season look for things that can be shipped easily, or buy them canned or frozen.

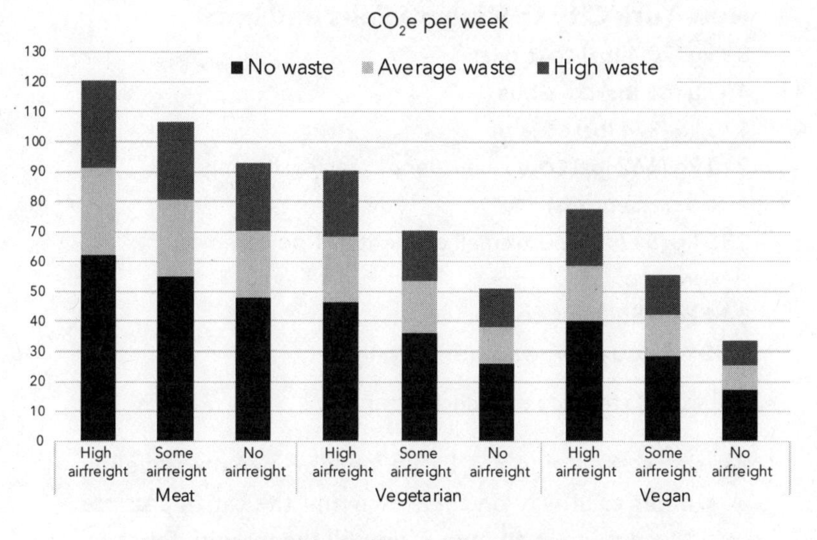

The weekly food footprints of different scenarios, based on diet, how much stuff is airfreighted, and waste habits

The other big variable is how much food you waste. According to a recent study, around 32 percent of the food bought in the US goes straight to landfill (see *What can we do?* endnote 3 for a discussion of this study). This rotting food will generate emissions, but far more important are the emissions from the unnecessary production of the food itself. If you waste more than average, then up goes your footprint as well as your shopping bill.

Other smaller things to think about are packaging; avoiding low-yield varieties; buying the wonky fruit and veg that is otherwise discarded, helping the store to reduce its waste; and how you cook it. (More details in *What can we do?*, p. 209.)

New York City to Niagara Falls and back

32 kg (70.4 lbs) CO_2e bike
40 kg (88 lbs) CO_2e bus
170 kg (374 lbs) CO_2e train
210 kg (462 lbs) CO_2e small electric car (driver only, no passengers)
235 kg (517 lbs) CO_2e small efficient gas-powered car (driver only)
344 kg (756.8 lbs) CO_2e plane
1.02 tons CO_2e large SUV (driver only)

> Still want to replace your car with an SUV?

All these scenarios are based on one person traveling 405 miles each way on their own and the car figures are based on driving at 70 mph. If you fill the car with four passengers, you can divide the emissions by four. This makes an electric car almost as competitive as taking a bus and easily beats train travel.

For all the road vehicles, the exhaust pipe emissions make up about half of the footprint. About one-third lies in the manufacture and maintenance of the vehicle itself. The supply chain of the fuel is responsible for the remaining one-sixth (see A *50-liter (13-gallon) tank of gas*, p. 129).

The bike is the outright winner if you can afford the time, you are careful about what fuel you eat (see *Cycling a mile*, p. 36), and you don't have a headwind against you. Of the more practical options, the bus comes out on top, with a footprint more than twenty-five times smaller than the gas guzzler. One reason that the bus beats the train is that buses travel more slowly, which is significant, because the energy needed to overcome air resistance goes with the square of the speed. Another reason is that, although a bus is heavy, the weight per passenger is much less than it is for a train (see *Traveling a mile by train*, p. 45).

Some analyses put a train ticket and a solo drive closer together in carbon terms. But I'm suspicious of these claims because the embodied emissions of the car per passenger mile are often ignored or underestimated. Whatever the precise difference (and it will of course vary widely depending on the particular vehicle), the train also lets you get some work done, read a book, or sleep instead of arriving at the other end stressed and frazzled.

The plane could actually be better than driving if you have the wrong kind of car (my sums are based on flying economy class). But please don't take this as an ad for flying—it's just a reminder of how carbon-profligate some road vehicles are.

As soon as there are more people on the trip, of course, cars become a lot more efficient. We used to load the whole family into my Peugeot 107, along with everything for a week's holiday, and put our bikes on the back (it was possible when the kids were younger). The emissions per passenger were lower than all of us traveling by train.

When it comes to both speed and safety, trains and planes win. When you are calculating how much of your life will be taken up by the journey, my back-of-the-envelope calculations tell me that a driver with a typical life expectancy should add about two hours each way to the car journey time to take account of the one in 200,000 chance that they will lose the rest of their life in a crash.[13] This is a very significant chunk to add on to the expected journey time of seven hours.[14] For trains and planes, the average loss of life expectancy through injury or death is tiny. I'm sad to have to report, for the sake of even-handedness, that the bike will lose hands down on safety grounds.

A common myth is that huge four-wheel-drive guzzlers are safer for their occupants. This is generally not true. They are, however, more dangerous for everyone else on the road.[15]

Using a smartphone

70 kg (154 lbs) CO_2e a year if you use your phone 1 hour a day
76 kg (167.2 lbs) CO_2e a year's typical usage of 3 hours and 15 minutes a day
96 kg (211.2 lbs) CO_2e a year if you use your phone 10 hours a day[16]
580 million tons CO_2e global mobile phone usage

> Typical mobile phone use connected to the internet works out at 1 g per minute, about the same as a large gulp of beer every hour[17]

Mobile phones are responsible for around 1 percent of global emissions but, for better or worse, typically take up 20 percent of our waking lives. The figures here include the electricity they use, their manufacture, and the networks and data centers they connect to.

Most of a smartphone's emissions come from its manufacture and transport to the user, in particular because of the precious metals and rare earths that need to be mined for smartphones' chips and motherboards and because people often replace their phone much earlier than necessary. It would take *thirty-four years* of average use for the footprint of the electricity you use to equal the footprint of the phone itself. So, if you keep your phone for twice as long, you almost halve the total annual footprint.

Apple has carried out detailed life-cycle carbon assessments of their products. They estimate emissions caused by manufacturing and transporting an iPhone 11 at 63 kg (138.6 lbs) CO_2e. However, the devil is in the details, and the life-cycle approach that they used has a nasty habit of "leaking"—missing out on little bits of the footprint. The footprint of a phone (or computer) comes from the complex mass of activity, starting with the mining of minerals. Behind

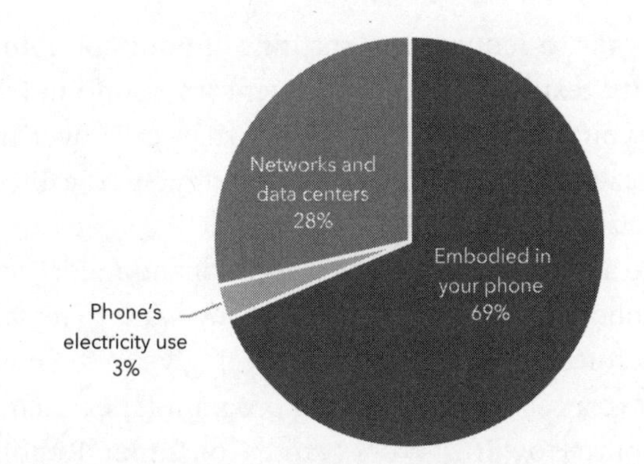

Carbon footprint of a year's smartphone usage (76 kg), including embodied and use-phase carbon, based on an iPhone 11 128 GB used for three hours and fifteen minutes a day over two years

each component lies a whole life cycle of processes that are almost impossible to map. You have to leave some processes out, thus cutting the pathways short, and the result is a shortfall in your footprint calculations that is known as "truncation error" (see p. 233 for more on this). In my figures, I have adjusted for this truncation error, which I estimate at 40 percent for the embodied carbon.

Thus adjusted, the embodied footprint of the iPhone 11 stands at 105 kg (231 lbs) CO_2e, which is 35 kg (77 lbs) per year if you keep it for three years, or 52.5 kg (115.5 lbs) if you discard it after two years as most people do.[18] However, keeping an iPhone for ten years would bring the embodied footprint down to 10.5 kg (23.1 lbs) per year. Factoring in power usage (4.2 kg CO_2e, or 9.2 lbs) and data transmissions (43 kg CO_2e, or 94.6 lbs), the total comes to 153 kg (336.6 lbs) over two years or 76 kg (167.2 lbs) a year.

In 2020, there were 7.7 billion mobile phones in use, with a total footprint of roughly 580 million tons CO_2e, which is roughly 1 percent of global emissions. This number is likely to increase, as there are still several billion people without smartphones.[19]

If you want to reduce the footprint of your communication habits, texting is a slightly lower-carbon option (see p. 21), and voice calls are lower carbon than calls over the internet because the networks use less energy. But the differences are relatively small.[20]

To make a real difference, buy a secondhand model, and keep your phone as long as possible.[21] If you want to replace a phone before its useful years are over, give it to someone else or to a cell phone recycling program (sites such as TechRepublic, HowStuffWorks, and Consumer Reports offer lists of recycling options).[22]

100 to 1,000 kilos (220 pounds to 1 ton)

A 50-liter (13-gallon) tank of gas

108 kg (238 lbs) CO_2e liquefied petroleum gas (LPG)
173 kg (381 lbs) CO_2e gasoline
196 kg (431.2 lbs) CO_2e diesel

> The US went through almost 143 billion gallons of gas in 2019[1]

The pie chart overleaf speaks for itself. If you were to pour a quart of gasoline on the floor and strike a match, the fumes would account for only about two-thirds of the carbon story. The other third is caused by the supply chain of the fuel: getting it out of the ground, flaring off the gas, shipping it around the world, refining it, and getting it to the pump. This extra third doesn't usually feature in car emission statistics (including official greenhouse gas conversion factors), which most often cover just the stuff that comes out of the exhaust pipe. This is one part of the reason why the carbon footprint of driving is often so badly underestimated.

Diesel has a slightly higher footprint per quart than gasoline (14 percent) but also has a proportionately higher energy content. Diesel engines are typically slightly more efficient at turning fuel energy into vehicle movement, so there is a small climate benefit in diesel cars over gas-powered vehicles. But there is also a colossal downside. Although they

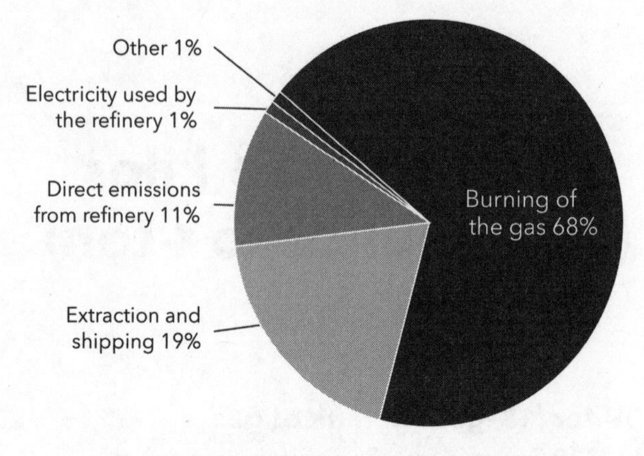

The carbon footprint of gasoline

have got a bit cleaner in recent years, diesel engines belch out many times more particulate pollution than gasoline engines, and this not only contributes to climate change (see *Black carbon*, p. 193) but is also estimated to result in tens of thousands of early deaths in the US every year. Diesel cars are bad news.

Biodiesel deserves a mention as a controversial option that is also full of problems. The first thing to say is that using land to grow fuel rather than food puts pressure on the world's forests, and chopping trees down already accounts for between 9 and 18 percent of global emissions (see *Deforestation*, p. 191). With a fast-growing world population, land is likely to feel increasingly scarce in future. The second negative point is that considerable emissions are involved in the growing of the fuel crop and the process of turning it into fuel. Some people think this can even add up to more than the emissions from fossil fuels.

On the plus side for biofuels is the potential to create them from unavoidable agricultural waste and the prospect that future technologies may be able to create them efficiently from algae. Overall, biofuels—if they are genuine sustainable fuels—might one day be a small part of the solution, especially for aviation, but they are not a magic wand.

A necklace

0 CO_2e handed down, or made from driftwood and shells
230 kg (506 lbs) CO_2e $700 worth of new Welsh gold
710 kg (1,565 lbs) CO_2e $700 worth of gold and diamonds from Africa

> Creating a single necklace can pump out as much CO_2e as a flight from Kansas to Kentucky

Who would have thought that something so small could have such an impact? But think about it for a moment and it makes sense: gold and diamonds are precious precisely because it takes effort, industry, and resources to extract them. And the carbon footprint comes almost entirely from their extraction, after which the actual forging of a necklace is relatively minor. To arrive at an overall ballpark figure for the carbon footprint of jewelry—460 g CO_2e per pound spent—I have again used the technique of working out the footprint of an industry and dividing it by that industry's total output.

At the bottom of my scale are items for which the value is in the art and not the materials, or a piece of jewelry that has been passed on or reforged from an existing item. The carbon impact here is simply from the energy required to melt it down and we can assume the footprint of extracting the gold or silver has been written off by previous owners.

For my average example, I have chosen a necklace of virgin Welsh gold. Although this has been mined especially for you, it will have been done using relatively efficient mining technologies and in a country where machinery tends to be more fuel-efficient than in low-income countries. At the top end of the scale is jewelry obtained using inefficient technology and cranky machinery. My figure of 710 kg (1,565 lbs) CO_2e is simply a crude estimate based on three times the carbon intensity of typical UK industry, which should be roughly the same for the US.[2]

While on the subject of gold and diamonds, be aware that chunks of the Amazon are being deforested in the pursuit of gold, often mined illegally,[3] and that workers in low-income countries are being exploited in the production of both gold and diamonds. Is it worth it? Can it really be a romantic gesture to give someone something that has an embodied footprint of exploitation? If you are feeling creative, there are lots of alternative possibilities, as per the zero CO_2e option above. If not, there are quite a few jewelry brands that claim to be environmentally sustainable and ethical.[4]

Christmas excess

4 kg (8.8 lbs) CO_2e per adult, low-carbon scenario
280 kg (616 lbs) CO_2e per adult, average
1.5 tons CO_2e per adult, high-carbon scenario

> A full-on Christmas could cost you four months' worth of 5-ton living

I said at the beginning that this book was about picking your battles. Christmas has got to be a good place to look, even if it might entail breaking a few habits and engaging in delicate family negotiations. For most of us there is a golden opportunity here to escape some mindless consumerism, stress, and perhaps even debt.

In my numbers I have only included excess: unwanted presents, wasted food, avoidable travel, lights for the house and the tree, and cards. This doesn't include the Christmas dinner consumed by you and your family. The numbers are *per adult* and are based on three scenarios, none of which is intended to be ridiculous.

The average adult spends about $660 on presents, of which 20 percent will be totally unwanted.[5] There will also be a lot of partly wanted middle ground, so I've assumed an average "wantedness" factor of 50 percent for all presents. In the festive season we spend about $300 more than usual on food, and I've allowed one-third for waste, thinking that this will be slightly higher than it is in the rest of the year because of the "Oh-no-not-turkey-again" effect and the fact that the big meals tend to keep coming over the whole period long after most of us have reached our wafer-thin-mint threshold. The Christmas lights burn through about 45 kWh. The average adult sends about twenty cards, with most of the footprint coming from their delivery. We typically travel 50 miles each above what we would do anyway, and it is generally by car.

In the high-carbon scenario, you spend $1,900 on presents (yes, that feels extreme to me too, but it's only a little over double the average). Sadly, in this scenario the wantedness factor turns out to be just 30 percent because you are even worse than me at choosing presents. People are too embarrassed to tell you or to sell them, so they gather dust or even get sneaked into landfill. You decorate your house with a wild lighting display that doesn't use LED bulbs. You mail 200 rather large cards. You also clock up 500 miles on a tour of relatives in a thirsty car.

I think the low-carbon scenario could be at least as festive and a lot less hassle. The food is great, but none gets wasted. You might eat a bit too much, but you make up for that over the coming months, so it's not additional. Your presents are thoughtful but not necessarily expensive. You encourage

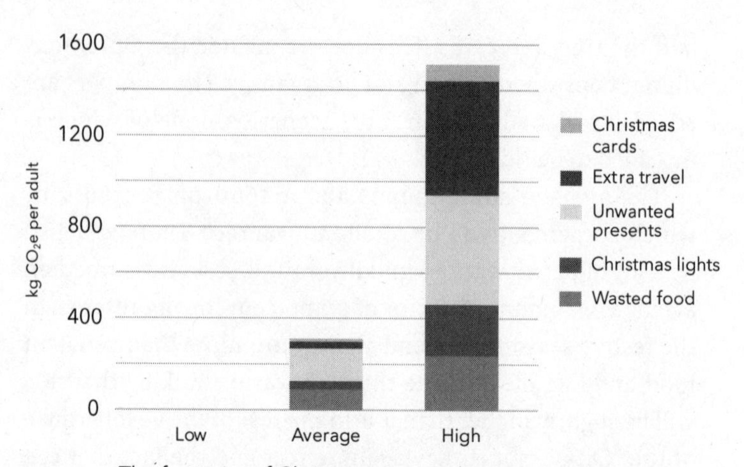

The footprint of Christmas waste–three scenarios

people to be honest in their reactions and you've kept all the receipts. You have LED lights for your home and tree. You stay at home and you send cards only to a few people that you haven't seen for ages and with whom you really don't want to lose touch. You video call your distant relatives and make plans to see them properly another time.

Some friends of ours decided that only children were going to get presents worth more than a strict limit of $1.50. They asked everyone to reciprocate, sending any cash saved off to the charity of their choice. Other friends decided that all family members would give just one present as a Secret Santa. Both giving and receiving became an exchange of gestures and altogether more fun.

A new carpet

76 kg (167 lbs) CO_2e thin polyurethane carpet with thin underlay, 13 ft × 13 ft

290 kg (638 lbs) CO_2e the same area covered in thick wool, polypropylene, or nylon with generous underlay

> If you lay 540 square feet of fine carpet, that could make up a fifth of your 5-ton budget

There is nothing intrinsically wrong with carpets. If you get full wear out of them, they may well pay for themselves in carbon terms by improving your insulation. However, if you are in the habit of moving every couple of years and insist on replacing everything with styles more to your own taste, then carpets, along with other soft furnishings, can be a significant chunk of your total carbon footprint.

The table below gives some figures for the footprints per kilo of fabric materials, based on European production.[6] In the UK, most textiles come from Asia, where industry is typically a lot more carbon intensive. For Chinese production, you can probably mark most of these numbers up by a factor of two or three on the basis that the factories tend to be

Carpet type	Carbon footprint (kg CO_2e/kg)
General	3.90
Felt underlay	0.97
Nylon	6.70
PET	5.56
Polypropylene	4.98
Polyurethane	3.76
Wool	5.53

Carbon footprints of different types of carpet[7]

less energy efficient and the electricity they use is also more carbon intensive per unit, as so much of it comes from coal-fired power stations. I'm not writing this out of some kind of protectionist instinct, just presenting the facts as I see them.

To give a sense of what the numbers mean in practice, typical weights are 0.2–0.3 lbs per square foot for underlay and 0.2–0.66 lbs per square foot for the carpet itself. This puts the overall footprint in the region of 1–3.7 lbs per square foot.

Insulating an attic

400 kg (880 lbs) CO_2e outlay for a detached house
(-) 40 tons CO_2e 40-year payback

> Insulating an attic can pay back the equivalent of eight years' worth of a 5-ton budget

Insulating an attic is the best carbon investment for anyone with a house. The embodied energy of the insulation material pays back in less than five months and is good for at least forty years. You will save about 40 tons of greenhouse gas in that period.

My calculations are based mainly on figures produced by the Energy Saving Trust[8] and assume you add 270 mm (10 inches) of Rockwool insulation to the previously uninsulated attic of a three-bedroom house. According to the EST's figures, you save 920 kg (2,024 lbs) CO_2e per year, but I've rounded this up to 1,012 kg (2,226 lbs) to take account of fuel supply chains they don't include.

In terms of money, even without a government grant, you'll get payback on your \$530 investment in just over three years. In fact, the decision to insulate your attic tomorrow will save you \$2,160 on top of paying back your outlay, compared with investing the money in a bank account with

	From no insulation to 270 mm (10 inches)	From 120 mm insulation to 270 mm (4.7-10 inches)
Cost without a grant	$530	$390
Annual payback	$305	$30
Embodied emissions in the material[9]	400 kg	224 kg
Annual carbon saving (including fuel supply chains)	1,012 kg	88 kg
Financial payback period (with 10 percent discount rate applied)	3 years	Never quite makes it
Payback over 40 years (with 10 percent discount rate applied)	$2,160	-$135
40-year carbon saving	40 tons	3 tons
Profit or cost per ton of carbon saved	Net profit of $54!	Net cost of $45

Insulating the attic in a three-bedroom house without a government grant

a 10 percent interest rate. It really is a no-brainer. In the US, you can get tax credits for insulating your home, among other ways of improving your energy efficiency.[10]

Our table gives a detailed breakdown for the scenario discussed and also for someone increasing their insulation from 120 mm to 270 mm (4.7–10 inches). This is a good move, too, if you care about the carbon savings or can get a grant. If you are just in it for the money, and you apply a discount rate, then I don't think you ever quite get it back again. However, at just $45 per ton, the CO_2e saved improving your existing insulation is still a hugely cost-effective way of investing in a lower-carbon world.

The EST's calculations that I've used are based on the assumption that, rather than cashing in on all the financial and carbon savings that would be possible, you will in fact allow your home to be warmer once it is insulated.

Various types of attic insulation are available: you can get the standard synthetic kinds as well as varieties from sheep's wool, paper, and a range of other options. Whichever you go for, make sure that you are 100 percent convinced there is no compromise on performance or longevity. Those are the two priorities.

A funeral

410 kg (902 lbs) CO_2e cremation
630 kg (1,386 lbs) CO_2e woodland burial
630 kg (1,386 lbs) CO_2e pyre
840 kg (1,848 lbs) CO_2e field burial

> Treat yourself to whichever method takes your fancy–after all, you will have just done the most carbon-friendly thing possible

In 2016, my company was commissioned to carry out something called The Corpse Project, a diverting interlude from our regular work crunching carbon numbers for corporates.

We assumed a funeral with fifty mourners traveling to the site of disposal in average cars (three people to a vehicle), and then a certain amount of close family visits. We added the fuel required for incineration. The woodland burial assumed a cardboard coffin, while the pyre included a cotton shroud. Although it turned out to be insignificant, we also took a look at the maintenance of the field.

Cremation wins because there is likely to be a crematorium near where you live, which means less travel (I've assumed the ashes stay on the mantelpiece or are scattered somewhere that would be visited anyway or not revisited at

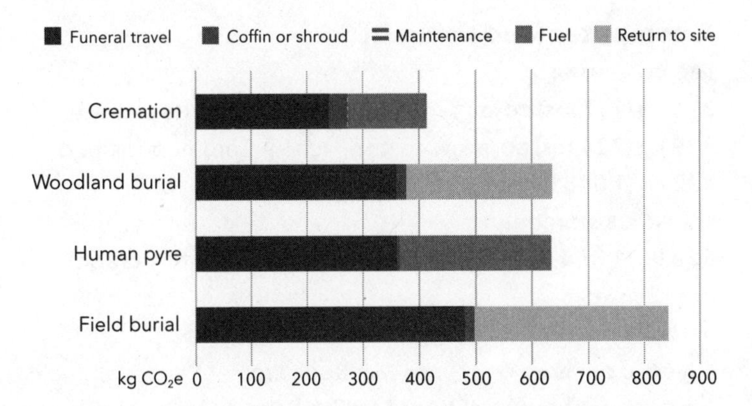

Four types of funeral, including mourners traveling to and from the funeral and return visits from closest loved ones

all). The field burial comes out worst because the site is more likely to involve longer journeys to get there. The pyre is an inefficient way of burning a body compared to a modern efficient crematorium furnace. I've assumed that the wood used could otherwise be used to replace fossil fuel.

An alternative option, sea burial, could be ideal if you died on a boat and no one makes a special journey out to see you go overboard. Over you go, with no combustion fumes and no transport footprint.

Traditional graveyard burial sounds like a more climate-friendly solution than a crematorium, but, browsing blogs on this subject (there really are people who write these) as I wrote the first edition of *Bananas*, I found a clergyman who reckoned that burials turned out 10 percent higher carbon than cremation once you take account of cemetery maintenance for the next fifty years.

To sum up, while this is one very special occasion when it really is okay to do whatever you want—after all, death is a full stop to your carbon footprint—if, for symbolic reasons, you do want to choose the low-carbon option, cremation probably wins.

A computer (and using it)

THE COMPUTER

326 kg (717 lbs) CO_2e 13-inch MacBook Pro, 128 GB storage
329 kg (724 lbs) CO_2e low cost 14-inch HP Chromebook 14 G5
475 kg (1,045 lbs) CO_2e 15-inch Dell Precision 5530, 256 GB storage
620 kg (1,364 lbs) CO_2e 16-inch MacBook Pro, 1 TB storage[11]

USING IT

4 g CO_2e per hour on 13-inch MacBook Pro
6 g CO_2e per hour on 16-inch MacBook Pro
12 g CO_2e per hour on 14-inch HP Chromebook 14 G5
20 g CO_2e per hour on average-efficient laptop
98 g CO_2e per hour on desktop computer with screen
130 g CO_2e per hour on gaming PC with screen[12]
+ 37 g CO_2e per hour for use of servers and networks[13]

> A new laptop has the footprint of a flight from Los Angeles to Seattle, but its use is low-carbon

Computers have fairly high footprints to their manufacture. HP, Dell, and Apple have all carried out detailed life-cycle carbon assessments of their products, but all miss out about 40 percent of the embodied carbon due to truncation error.[14] Apple's website talks about reducing its impact by making machines lighter and using renewable energy in their aluminum production. All this helps, but the bulk raw materials are just a small part of the issue. If a laptop were just a lump of plastic, steel, and semiconductor, you could get its footprint to something like 10 kg (22 lbs) CO_2e. The problem is that microprocessors come in at around 5 kg (11 lbs) CO_2e for a 2 g chip, because of the high-tech process that is involved and the incredibly clean environment required.[15] Apple also talks about reducing packaging; this, too, is good practice but makes a marginal difference.

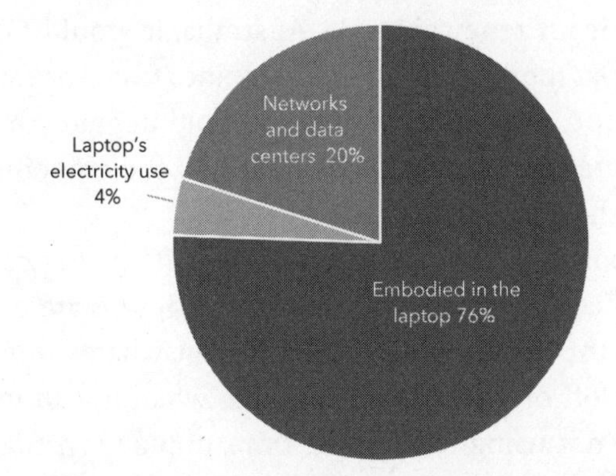

A year's use of a 13-inch MacBook Pro you keep for four years and use for three hours a day (including embodied and use-phase carbon and electricity from networks and data centers)

It's hard to give guidance on what makes a low-carbon computer, because the processes involved in making one are so complex. The guidance we would get from input-output analysis is that the carbon footprint is proportional to the price of the machine. This is almost certainly not right. As a rule of thumb, I would expect higher-quality, more expensive machines to have a slightly higher footprint but to be better carbon value in terms of the use we ought to be able to get out of them as long as we make them last. The most important thing you can do is to get many years of life from your machine. Get it repaired if it breaks and pass it on if and when you need to upgrade.

As to actually using a computer, this is a fairly low-carbon way of spending time. Apple reports that the 13-inch MacBook Pro consumes 5.8 watts of electricity when in use and that the power supply is 90 percent efficient. So that makes a total of 6.4 watts leaving your mains plug. My estimate, therefore, is that in the US the emissions from its use are 4 g CO_2e per hour. This would equal the machine's embodied emissions after 80,000 hours in the US. (In the UK, it would take 150,000 hours, as less electricity comes from coal and more comes from renewables; in Australia, it would take 60,000 hours, as more of the electricity comes from coal; and in China 50,000 hours.) Most people would upgrade their machine before clocking up those hours, so the embodied emissions in the machine are the biggest deal.

When choosing a computer, think about its power consumption, especially if you will use it a lot. Laptops are usually better in this respect than desktops, but whatever you use, switch it off or put it in sleep mode when not in use. And if you're not using it for some time, unplug it if that's what it takes to zero the power.

I haven't included the use of peripherals or the activity you might stimulate in other machines around the web through your emails and web searches (see *The Cloud and the world's data centers*, p. 179).

A pet

25 kg (55 lbs) CO_2e per year for a goldfish[16]
310 kg (682 lbs) CO_2e per year for an average-sized cat
770 kg (1,694 lbs) CO_2e per year for an average-sized dog[17]
2,500 kg (5,500 lbs) CO_2e per year for a Great Dane

> A couple of Great Danes could use up the whole of your 5-ton lifestyle

I'm not writing this to give you a hard time about having a pet, which I know can be a huge part of any household. We've even ended up with a dog in my family, and although I'm still not totally sure how I let that happen, as a traditional non-dog-lover, even I have to admit that there is a plus side.

The footprint of dogs and cats largely boils down to the food they eat, and it doesn't help that they are both carnivores. The web is full of hot debate on whether you can feed a dog a vegetarian diet, which would probably cut its footprint to about a third. As far as I can make out, it looks risky to their well-being, with a major problem being that their short digestive tracts aren't cut out for most vegetables (writing this will, I know, prompt a surge in the *Bananas* inbox).[18] However, you can still cut the footprint a long way if you opt for chicken rather than beef and don't overfeed them.

The size of your pet also makes a huge difference, as a big dog will eat up to ten times as much food as either a small dog or an average-sized cat.[19] Being vegetarians, rabbits have dramatically lower footprints than even the smallest dogs. Gerbils, mice, and rats even lower still. Goldfish are hardly on the scales.

For the numbers above, you can also add something like 11 kg (24.4 lbs) CO_2e for every $100 you spend at the vet (and, from experience, don't underestimate what you could be letting yourself in for).[20]

A couple more statistics before leaving the subject of pets. America's dogs are estimated to produce a massive 5.1 million tons of poo per year.[21] That also means tens of billions of plastic bags—very nasty unless of the biodegradable type. Meanwhile, the UK's 8 million cats have been estimated to kill 200 million birds and small mammals per year, so the average cat makes about one kill every two weeks.[22]

On the positive side, all pets make it harder for you to fly off on holiday, so they may end up saving carbon in the long run. And, if you have a pet rather than a child, that is a huge carbon saving.

A mortgage

1,000 kg (2,200 lbs) CO_2e per year $270,000 on 2.5 percent interest

> How can a mortgage have a carbon footprint?

How indeed? Surely a mortgage just boils down to a few bits of paper or electronic transactions? Look more closely, though. The bank or mortgage company runs offices, buys computers, sends mail (probably mainly junk mail), and stores data. Its employees travel. It outsources everything from cleaners to building maintenance, from design work to corporate lunches, and maybe even still buys the odd paper clip.

What I am saying is that when you take out the loan you feed the financial services industry along with all its direct and indirect environmental impacts. This is another example of a set of ripple effects across the economy that we can't see and don't stand a chance of counting on an individual basis. Happily, our input-output model (see p. 234) comes to the rescue and produces a ballpark figure of 141 g CO_2e for every dollar spent in the US.

If you have a $270,000 mortgage on a 2.5 percent interest rate, you pay $6,750 per year (plus any actual repayments) and this incurs an annual footprint of the order of 1,000 kg (2,200 lbs) CO_2e. The same story applies to all loans, and the principle goes wider still. All the intangible services have fairly similar carbon intensities: solicitors, lawyers, accountants, therapists, architects, and so on.

There are two basic lines of attack if you want to cut the carbon. The first is to take out a smaller mortgage and spend the money you saved on something that decreases carbon emissions, such as an investment in an offshore wind farm, a save-the-rainforests project, or (perhaps best of all) a solar roof (see *Solar panels*, p. 154). You could stick the money in the bank where it may seem harmless, but even then you may be enabling the bank to lend more to profligate consumers.

The other line of attack is to be discerning about the way the mortgage company goes about its affairs. I have based the footprint estimate on general figures for the industry, but there are good and bad practices within it. Ten years ago, one-tenth of the sector's footprint came from printing and postage, but this is now only 2 percent. A much bigger deal is the 35 percent that comes from air transport.

However, there are some good examples. The Ecology Building Society runs a simple, lean operation out of eco-friendly premises and makes a real effort to walk the talk. My company estimated their emissions intensity to be about a quarter of the industry average.[23] Furthermore, their footprint is in the cause of encouraging a sustainable building stock, because they vet their loans by the sustainability of the project and also support lenders in improving their buildings.

More widely, every time we invest or allow a bank or pension plan to do so on our behalf, we are pushing for one kind of future or another.

On the one hand, banks vary considerably in how much they invest in fossil-fuel industries. According to Oil Change International, JPMorgan Chase, Wells Fargo, Citi, and Bank of America in the US and Barclays, HSBC, Santander, and RBS/NatWest in the UK score pretty badly. On the other hand, in the US, Mighty Deposits offers a list of banks and credit unions that operate sustainably and have made commitments not to invest in fossil fuels.[24]

1 to 10 tons

An operation

1 ton CO_2e hip replacement or knee surgery
2.3 tons CO_2e heart bypass operation
31 million tons CO_2e UK healthcare total
480 million tons CO_2e US healthcare total

> Healthcare accounts for around 4 percent of the UK's overall emissions—and as much as 8 percent in the US

Let's start with the UK. The carbon cost of healthcare averages around 180 g CO_2e per pound spent.[1] That makes it a fairly low-carbon way of spending money. And in terms of the quality-of-life improvements we stand to gain from it, healthcare when we need it must be one of best ways of spending our carbon budget.

That said, a big operation clocks up a big footprint. The typical cost of a heart bypass to the National Health Service in the UK is about £13,000 ($17,550).[2] If we make the very crude assumption that this operation is averagely carbon intensive, that adds up to 2.3 tons CO_2e. This is higher than some of the more common operations, like major hip or knee surgery, which aren't quite as lifesaving but only come in at around 1 ton per surgery.[3]

Overall, my best estimate for the footprint of UK healthcare, including all pharmaceuticals, was 31 million tons CO_2e

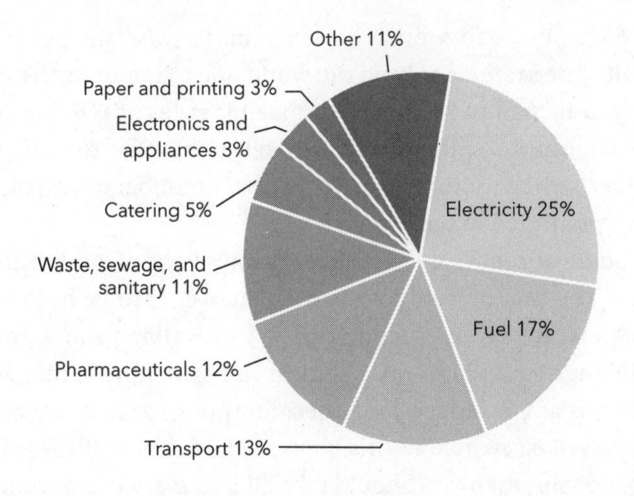

The carbon footprint of UK healthcare (2018)

in 2018, or just over 4 percent of the UK's total footprint.[4] (By comparison, healthcare was about 9.6 percent of the country's GDP.) Electricity and fuel accounted for 42 percent of this figure, transport for 13 percent, drugs for 12 percent, and everything involved in keeping the hospitals and practices clean 11 percent. Paper and cardboard accounted for a massive 3 percent of the footprint of all healthcare despite the rise of IT, which along with other electronics and appliances also accounts for about 3 percent. I'd like to think that the paper is not the stuff that clogs up the filing cabinets of one of the world's biggest bureaucracies, but rather the consumables used to keep things clean.

For the US, the overall healthcare footprint is 480 million tons CO_2e, according to a recent paper, which equates to 1.5 tons CO_2e per capita.[5] That's compared with .5 tons CO_2e per capita in the UK. The higher healthcare footprint per capita can be mostly explained by a higher carbon intensity of the US energy system; however, the higher healthcare expenditure also has a significant effect.

Sadly, in 2020 we have seen a massive strain put on healthcare systems around the world due to the outbreak of COVID-19. As I write this, the pandemic rages on. On top of the huge losses of life, this is also likely to have led to a much higher carbon footprint due to the sheer number of patients being admitted to the hospital.

So, what can we do to reduce the emissions of our healthcare? The best option, now more than ever, is to be healthy. This might involve cycling (safely) or walking more and thinking about the amount of meat and dairy in your diet, all things that will reduce your direct footprint, too, and that are discussed elsewhere in this book. And, of course, if there is a pandemic, then you should take all necessary precautions to prevent yourself getting infected. The footprint of a mask and hand sanitizer is many times smaller than a lengthy stay in the hospital.

When you do actually need healthcare, be as careful as possible with resources (even if you have good health insurance). But relax in the knowledge that it is one of the lower-carbon ways for you to spend money.

A ton of steel

400 kg (880 lbs) CO_2e general steel, 100 percent recycled
1.1 tons CO_2e virgin steel, using hydrogen instead of coal in the manufacturing process
1.8 tons CO_2e global average, 25 percent recycled
2.3 tons CO_2e virgin steel, blast furnace with coal
2.9 billion tons CO_2e annual global production[6]

> Steel makes up 8 percent of the world's CO_2 emissions, and just 25 percent is recycled

The figures above are for a ton of steel in its raw form at the foundry gate. In other words, they do not take account of any additional emissions that might be required to transport it to wherever it will be used or turn it into something useful like a car or a part of a house. The value of recycling immediately becomes evident, because recycled steel has one-fifth of the footprint of its traditionally made virgin equivalent.

As the world decarbonizes, the steel industry is going to have to play its part. Here are the four key things that need to happen:

- **The whole world needs to dematerialize** We need fewer cars, and to make them last longer. We need to buy fewer but higher-quality goods, pre-loved if we can, make them last, and get them repaired instead of chucking them out when they break. When we've finished with them, we need to sell them or pass them on.
- **We need to use sustainable materials where we can** About a third of all steel is used in the construction industry[7] and, while it is hard to build a skyscraper out of wood, it is perfectly possible to build wooden houses.
- **We need to increase recycling rates** These have to go a long way from today's pitiful 25 percent global average.
- **We need industry to shift to using hydrogen instead of coke (which comes from coal) as the reducing agent needed to extract iron from ore** This switch would reduce steel's carbon footprint by half. The technology for this is ready to be deployed at scale right now and at the moment it is perhaps only 10 percent more expensive than a traditional blast furnace. In other words, all we'd need is a very modest carbon price of about $35 per ton of CO_2 to make this the cheapest method.[8] Or a government that is willing to invest.

And here is one thing that emphatically does not need to happen:

- **We do not need to open up new coal mines for coking coal extraction** This topic is close to my heart. I could scream with frustration at the madness of my own county council approving one of these on my doorstep in the Lake District. It has taken me a long time to understand how such a poor decision could have got so far. I think it boils down to a mix of willful ignorance and timidity on the part of councilors, planning officers, and the Secretary of State, and the fact that guidelines for environmental impact assessments date back to an age when climate was treated as a side issue (see *A new coal mine*, p. 176).

A ton of nitrogen fertilizer

2.7 tons CO_2e nitrogen fertilizer efficiently made and sparingly spread

12.3 tons CO_2e the same stuff made inefficiently and used in excess

> By being more careful with fertilizer use, there is a real carbon-saving opportunity

Nitrogen fertilizer is a significant contributor to the world's carbon footprint. Its production is energy intensive because the chemical process requires both heat and pressure. Depending on the efficiency of the factory, making 1 ton of fertilizer creates between 1 and 4 tons CO_2e. When the fertilizer is actually applied, between 1 and 5 percent of the nitrogen it contains is released as nitrous oxide (which is around 300 times more potent than CO_2). This adds 1.7–8.3 tons CO_2e to the total footprint (the wide range here depends on a variety of factors).[9]

Here's how the science of it goes. All plants take up nitrogen, so if you're growing a crop, it has to be replaced in the soil somehow or it will eventually run out. Nitrogen fertilizer is one way of doing this. Manure is another. There can be big benefits from using fertilizer on crops. For some crops, in some situations, the amount of produce can even be proportional to the amount of nitrogen that is used. However, there is a cut-off beyond which more nitrogen does nothing at all to the yield, or even decreases it. Timing matters, too. It is inefficient to apply fertilizer before a seed has had a chance to develop into a rapidly growing plant.

China is the world's largest consumer of nitrogen fertilizers, followed by India and the US. Some 23 percent of all nitrogen fertilizer is applied to fields in China—about 25 million tons per year[10]—and Chinese farmers use more than four times as much fertilizer per acre as the global average.[11] Many have a visceral sense of the need for high yields, having experienced hunger in their own lifetime, so it is easy to understand the instinct to spread on a bit more, especially if encouraged by the fertilizer industry. After all, China has 18 percent of the world's population to feed, from 8 percent of the world's arable land.[12]

The good news is that there has been progress. An advice program to 21 million Chinese farmers led to an 11 percent cut in fertilizer use and a 16 percent increase in yield[13] between 2005 and 2015. Many other countries could benefit from a similar initiative.

But not all fertilizer use is bad. In parts of Africa, for example, there is a scarcity of nitrogen in the soil and there would be real benefits to applying a bit more fertilizer to increase the yield and get people properly fed.

Flying from Los Angeles to Barcelona return

3.5 tons CO_2e economy class
4.5 tons CO_2e premium economy
10 tons CO_2e business class
14 tons CO_2e first class

> One business-class flight from Los Angeles to Barcelona and back is twice the year's 5-ton budget (or the equivalent of 340,000 plastic bags)

A Boeing 747 carrying 416 passengers burns through 116 tons of fuel on the 9,700 km (6,030 mile) flight from Los Angeles to Barcelona. Almost one-third of its total weight on take-off is fuel. As the fuel burns, it creates three times its weight in CO_2.

However, the impact is even worse than that because high-altitude emissions are known to have a considerably greater impact than their low-altitude equivalents. The science of this is hideously complex and poorly understood,[14] but there is a clear case for applying a multiplier to aviation emissions to take account of their extra impact. I have used a factor of 1.9 (see *A brief guide to carbon footprints*, p. 6).

Aviation is sometimes said to account for only about 2 percent of global emissions. These statistics ignore the effect of altitude and are much, much higher in the developed world.

Your own aviation footprint may well be much less than this. Many Americans never fly at all. By contrast, for some people, flying accounts for the overwhelming majority of their total footprint, and trying to cut carbon in other areas might make almost no difference at all. First-class and business-class tickets are particularly high in impact, simply because your seat uses up more of the plane and because by paying more money you provide a greater proportion of the commercial incentive for the flight.

It's hard to imagine a low-carbon flying technology coming to the rescue anytime soon. Electric planes do exist, but they are in their infancy, restricted to a few seaplanes in Canada and experimental craft elsewhere. The physics of flight simply does not allow us to reduce the energy it takes to keep us in the air by more than a few percent, and for the foreseeable future that energy has to come from fossil fuels. Nevertheless, there are still some efficiencies to be had. One of these is the automation of air-traffic control to calculate optimum flight paths, which some estimate could bring in efficiency improvements of 9 percent.

Ultimately, though, it's hard to avoid the conclusion that we need to fly less. That needn't make our lives any worse. Make your flights count: go for longer but less often and do things you really couldn't do at home. For the rest, try local trips, which involve less travel time and more holiday. After all, the experience of getting to an airport, hanging around in a departure lounge, and then sitting cooped up for hours is a rubbish way of spending time. Also think about *where* you fly to: the closer the destination, the fewer the emissions. One myth is that long-haul flights are automatically more efficient per mile than short-haul flights because they involve proportionally less time taxiing, queuing, taking off, and landing. This isn't necessarily true, because the long-haul flight has to lift more fuel. The most carbon-efficient way of getting across the world is in several hops—but not too many. But none of this changes the fact that the farther you fly, the larger the footprint.

Of course, the flying conundrum affects companies as well as individuals. I work with a few businesses for whom flying is a key issue. They know it's high in carbon, costly, and time-consuming, but they thought they had strong business reasons for doing it. Perhaps COVID-19 will have delivered a permanent shot in the arm, with the explosion of Zoom calls and other video conferencing.

It is difficult to see a place in the low-carbon world for much airfreighted food (see A 250 g (8 oz) *bunch of asparagus*, p. 86), let alone durable goods such as clothing. Some garments are airfreighted simply to reduce lead times and cut the cost of stock that is tied up in transit at sea. Airfreight labels are one piece of consumer information that would surely be simple and helpful. Currently, these are found on some supermarket fresh produce but nowhere else.

I'm sometimes asked about airfreight from developing countries: "Surely it's good to keep supporting that country by carrying on the trade!" In broad terms, I don't think so. The argument is a bit like saying you should keep the arms trade booming so that people can keep their jobs. Economies need to be powered by people doing things that are useful. Anything else is unsustainable nonsense. And it is amazing how often exports involve a country, in effect, selling its (often much needed) water, which is embodied in everything from cotton to avocados.

Solar panels

5 tons CO_2e for a 4 kW domestic array

(-) 2.5 tons CO_2e annual carbon saving from a 4 kW domestic array

> Installed in the US, solar panels should become carbon-negative within two years

The carbon footprint of solar power is made up of the embodied carbon in the manufacture of the panels and other kit. And that in turn depends on carbon intensity of the energy used to manufacture them as well as the amount of energy required. A summary of studies from 2011[15] arrived at a figure of 2.5 tons CO_2e for the manufacture of a 1 kWh array of panels, but since then the manufacturing process has steadily increased its mix of renewable electricity and panels have become more efficient at turning sunlight into energy. On that basis I have halved my estimate compared to the 2011 figure. My numbers also assume only average US-quality sunlight and a twenty-five-year lifespan for the panels; you may well get longer use from them.[16]

The size of the array refers to the peak output in the sunshine. I've based my calculations on a US typical capacity factor of 12 percent, which is the average power spread across all times of day and night through the year, compared to the peak output. So, a 4 kW array averages 480 watts and collects 4,200 kWh per year. In a really sunny part of the world, you might get the capacity up to close to 1 kilowatt (about 24 percent), doubling the annual output to around 8,500 kWh and halving the carbon intensity of the electricity.

As of mid-2020, US grid electricity comes in at 650 g CO_2e per kWh (once you take account of the emissions in its supply chain pathways), so the carbon savings of your solar panels work out to 600 g per kWh. That makes an annual saving for a 4 kW domestic array of around 2.5 tons CO_2e, and thus a carbon payback period of about two years for installing it.

Over its lifetime, the 4 kW solar array will generate about 104,000 kWh in the UK, which is more than ten times the energy required to make them.[17] Based on today's installation costs for panels of $2,190 per kilowatt, this means you are paying about $140 per ton of carbon saved. (Interestingly, as grid electricity decarbonizes, so does the carbon benefit of your solar array. In time, all domestic gas used for heat will need to be replaced with renewable electricity.[18])

However, you'll also save money from the electricity that the panels generate, and, in purely financial terms, you can expect your panels to pay back in about ten years, based on a saving of 10 cents per kWh. (Some of the time you will save the full cost of electricity that you would have otherwise had to purchase, some of the time you'll save just the cost of the gas you might have used instead to get heat, and sometimes you will just be getting the small payment your electricity provider will give you to feed it back into the grid. So 10 cents per kWh is a rough average.)

Although solar energy is a very low-carbon way of producing power, its footprint is still significant enough to be a consideration for policy makers. At an absolute minimum, I think we are going to need to generate about 85,000 TWh (1,000,000,000 kWh) of electricity per year. At today's carbon footprint per solar panel, that would work out to 60 billion tons CO_2e, or more than a year's worth of global greenhouse gas emissions. Luckily, going forward solar panels are likely to require fewer raw materials, so the embodied energy and carbon per panel is likely to fall to half or less of today's levels.

Already prices for domestic solar installations have come down to the point that, while not necessarily the most lucrative way of spending your money, you only have to care about the environment a little bit for them to make sense.

There is an easier calculation for larger-scale solar farms, where both the initial costs and the carbon are less per kilowatt-hour.

Finally, remember that solar power doesn't do us any good unless we have it instead of the fossil fuel, rather than additionally. The world needs to stop growing its energy consumption in order to give the new renewables a chance to replace the fossil fuel.

10 to 1,000 tons

A new car

4 tons CO_2e Citroën C1, Peugeot 107, basic specs
8 tons CO_2e Ford Focus Titanium
11 tons CO_2e Renault Zoe (electric)
12 tons CO_2e Toyota Prius plug-in hybrid
25 tons CO_2e Range Rover Sport HSE

> A Range Rover or SUV could eat up ten years of a 5-ton lifestyle before you even drive it

The carbon footprint of a car is complex. Ores have to be dug out of the ground and the metals extracted. These have to be turned into parts. Other components have to be brought together: rubber tires, plastic dashboards, paint, and so on. All of this involves transporting things around the world. The whole lot then has to be assembled, and every stage in the process requires energy. The companies that make cars have offices and other infrastructure with their own carbon footprints, which we somehow need to allocate proportionately to the cars that are made. The ripples go right through the economy.

Attempts to capture all these stages by adding them up individually (the so-called process-based approach to carbon footprinting) are doomed to result in an underestimate,

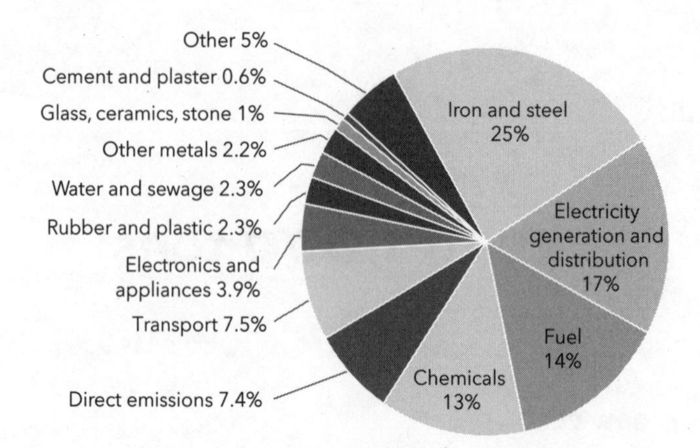

The very complex carbon footprint of a new car (2016)

because the task is just too big. Luckily, there's an alternative in the form of the input-output method (see p. 234). This approach takes account of all these infinite ripples, even if it does rely heavily on the law of averages. It can give us clues as to the footprint of a car per unit of monetary value and also tell us a bit about how that footprint comes about.

The input-output approach suggests that a car might have a footprint of 267 kg (587 lbs) CO_2e for every $1,000 that you spend on it. The pie chart above shows how this breaks down, and its complexity illustrates just how far and wide the footprint is dispersed. There is only room to put labels on the biggest slices: the manufacture of iron steel makes up a quarter; electricity and fuel (including gasoline and coal) make up 30 percent; chemicals are also significant. The rest of the footprint is dispersed far and wide, and our pie is just the beginning, for behind each piece are all the complex supply chains supporting each industry sector.

The upshot is that the embodied emissions of a car typically rival the exhaust emissions over its entire lifetime. And the range between models is vast. Manufacture of a top-of-the-line Range Rover Sport HSE that ends up being scrapped after 100,000 miles accounts for about two-thirds of its lifetime exhaust pipe emissions. For my much smaller Peugeot 107, it would be just one-third, even though the 107 only burns about one-third of the fuel per mile.

I have seen plenty of analyses of whether it is a lower-carbon option to keep or to scrap your old car. These almost always rely on process-based approaches and therefore underestimate the embodied energy and conclude that you should replace your car far too readily. Even as we transition to electric cars, it doesn't make sense to get rid of your old fossil-fuel-powered car before its time, unless it is one of those gas guzzlers that should never have been built in the first place.

Generally speaking, then, it makes sense to keep your old car for as long as it is reliable, unless you are doing high mileage, or the fuel consumption is ridiculously poor. You can, of course, boost the life of the car by looking after it. The table overleaf shows how much lower the total emissions per mile can be if your car lasts twice as long.

If your old car does reach the end of its days before you start thinking of replacing it, take a look into ride-sharing programs: you may save a lot of money as well as reduce the number of cars that need to be produced. If you decide you really need a new car, opt for a secondhand, light, simple, and fuel-efficient model (that way you'll be limiting both the manufacturing and the exhaust pipe emissions) or an electric car. If you are buying new, get an electric one—the smaller, the better.

Electric cars do away with the footprint of an internal combustion engine but swap that for the even higher carbon footprint of a battery; so, like-for-like embodied carbon is significantly higher. However, the carbon footprint of the

Model	Price in thousands of pounds (thousands of US dollars)	Embodied emissions (tons CO_2e)	Embodied emissions per mile over 100,000 (grams CO_2e)	Embodied emissions per mile over 200,000 (grams CO_2e)
Citroën C1	£11-15 ($15-21)	4-5	39-54	20-27
Ford Focus Titanium	£23-27 ($32.5-38)	8-10	82-99	41-49
Renault Zoe (electric)	£27-31 ($38-44)	10-11	98-114	49-57
Toyota Prius plug-in hybrid	£32-34 ($45-48)	12	117-124	58-62
Range Rover Sport HSE	£69-71 ($97.5-100)	25-26	249-259	124-129

The carbon footprint of cars per mile of use[1]

electricity you use is a lot lower than the fossil-fuel alternative per mile (more so in the UK with its renewables than, for example, Australia or the US). Overall, electric cars are about twice as carbon friendly as gas-powered cars.

I am more pessimistic about electric cars than some analysts, as I think the embodied carbon in manufacture is almost always underestimated.[2] Plug-in hybrids could be a good option if you want to go electric but need the range of a conventional car and don't want a huge battery. Hybrids that you can't plug in (like the old Priuses) get all their energy from fossil fuel, with only a fairly small efficiency advantage from regenerative braking.

A person (annual footprint)

0.2 tons CO_2e average Malawian
7 tons CO_2e world average
8 tons CO_2e average Chinese
13 tons CO_2e average Briton
20 tons CO_2e average Australian
21 tons CO_2e average North American

> We need to assess our footprints not just within our own borders but for all we consume

At 21 tons, the average US person has one of the biggest footprints of anyone in the world, not least because they are richer than average. The good news is that it has come down slightly, from about 28 tons per person ten years ago. In many respects, though, not much has really changed. And it needs to. The 5-ton lifestyle is an urgent intermediate step on the way to global net zero.

I don't know if you can account for the higher footprint entirely because the US is a wealthier country overall. There is such enormous inequality in the US that many aren't wealthy at all. (I go into more detail on this in one of my other books, *There Is No Planet B*.) There are other factors. For example, in the US, many needlessly inefficient vehicles burn through a lot of very cheap fuel. Is that a function of wealth alone? A lot of US electricity still comes from coal, relatively little from renewable sources. Is that rooted in affluence or are there other factors? American homes tend to be larger than in many other nations, which means more fuel use. Is that purely a function of affluence, or is there more to it? The average Chinese person's footprint has more than doubled in the ten years since the first edition of this book. Surely, that has something to do with rapidly increasing affluence, but also that there's still plenty of coal in the electricity mix, and probably other factors too. Regardless, it's sobering to realize

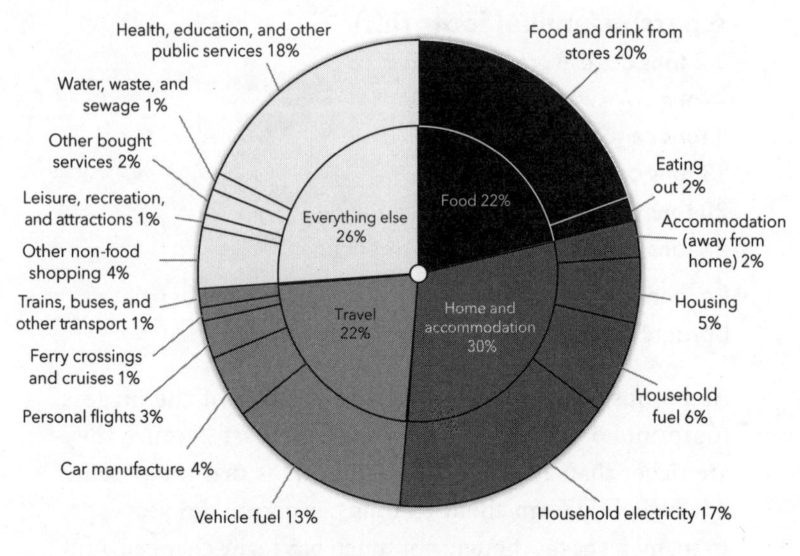

Footprint of an average person in the US

that it takes about 105 Malawians to make up the footprint of just one US person.

My estimate for the average US person's 21-ton carbon footprint includes everything we do and buy. It divides up neatly into four parts:

- **Food**—including far too much meat, dairy, waste, and airfreight.
- **Homes**—building new homes, maintaining old ones, gas and electricity bills, and also all accommodation away from home.
- **Travel**—mainly cars and air travel.
- **Everything else**—basically all the inedible things you buy other than cars and houses, as well as services of every kind, including your contribution to the running of all national services from healthcare to the police, military, and government.

So, if you are trying to cut your footprint down, you can think about it in four nice, neat chunks. In the chapter *What can we do?* (p. 209), I've gone into quite a lot more detail on this.

Space tourism and travel

330 tons CO_2e Blue Origin New Shepard tourism flight
600 tons CO_2e SpaceX Falcon 9 flight to the International Space Station
7,060 tons CO_2e space shuttle flight
10,300 tons CO_2e first flight to the moon, 1969

> Space tourism is truly the worst idea Bezos, Branson, and Musk have ever had

Space rockets use an enormous amount of fuel, mainly for taking off, when they need the most thrust. Modern spacecraft mostly use a refined form of kerosene and cryogenic liquefied oxygen or hydrogen peroxide.[3] The fuel consumption is determined by a few factors, including the weight of the spacecraft, the efficiency of the engine, and how high above the Earth it's trying to go.

Blue Origin's New Shepard, the private space tourism initiative of Amazon owner Jeff Bezos, aims to take space tourists just 100 kilometers up into the atmosphere and return within ten minutes. This is estimated to consume only 5.9 tons of kerosene and 49 tons of hydrogen peroxide per launch and therefore has a footprint of "only" 330 tons for its pilot and three passengers (110 tons per passenger).[4] That's over two years of 5-ton living *per minute.*

Rival SpaceX, the plaything of Tesla chief Elon Musk, also has plans for space tourism. Meantime, it is sending its Falcon 9 craft to travel 400 kilometers to the International Space Station. These flights go through 147 tons of kerosene and 341 tons of liquid oxygen.[5] By comparison, NASA's space shuttle, which was retired in 2011, burned up 103 tons of hydrogen, 616 tons of liquid oxygen, and 1,000 tons of extra high-energy solid fuel.[6]

At the top end of the scale, NASA's Saturn V Moon rocket, which brought Neil Armstrong and Buzz Aldrin into space in 1969, burned through 616 tons of kerosene, 17,000 tons of liquid hydrogen, and 1,390 tons of liquid oxygen, and delivered the "giant leap for mankind" with a carbon footprint of just over 10,000 tons CO_2e.[7]

My carbon estimates for these space flights are conservative. I have only factored in emissions from the fuel consumed per flight, with no weighting factor to take account of the high altitude at which the emissions are released. For planes, it's common to apply a markup of 1.9, as greenhouse gases have a far higher impact in high altitudes than they do at ground level. But since spacecraft use different fuels and at different altitudes, the impact on global warming is less clear.

I have assumed that the process of creating the hydrogen and solid fuel using energy from fossil fuels has been 80 percent efficient—that is, that four-fifths of the energy in the fossil fuel is transferred into the shuttle fuel. That is about as efficient as hydrogen generation ever gets. Much more significantly, it might have been reasonable to add on a large chunk of footprint from NASA itself. For the space shuttle, Richard Feynman, the Nobel Prize–winning physicist who helped to investigate the Challenger disaster, describes the whole project as NASA's somewhat unjustifiable raison d'être after the lunar landings.[8]

Finally, I haven't factored in the embodied energy of the vehicles themselves. This is probably a modest component, though, as each shuttle (apart from Challenger, which crashed after ten trips) was reused around thirty times.

A wind turbine

30 tons CO_2e 15 kW turbine, installed
(-) 500 tons CO_2e 15 kW turbine, 20-year payback
134 tons CO_2e 100 kW turbine, installed
(-) 2,619 tons CO_2e 100 kW turbine, 20-year payback
1,046 tons CO_2e 3 MW turbine, installed
(-) 81,538 tons CO_2e 3 MW turbine, 20-year payback

> Wind power is amazing at saving carbon, and the bigger the turbine the better

How does a wind turbine stack up in both cash and carbon terms? The answers vary considerably according to their size. At the lower end of the scale are micro-renewables. These are not necessarily small. A large and efficient one—a 15 kW turbine with a 9 m diameter and a pole as high as a four-story house—would cost around $94,500 to purchase and install and would be capable of delivering 36,000 kWh of electricity each year if placed on a suitably windy site. If turbine manufacture is about as carbon intensive per dollar of product as other generators and electrical motors, which seems a reasonable assumption, its carbon intensity will be around 300 kg (660 lbs) CO_2e per $1,000 of value. That makes the footprint of the installed turbine 30 tons CO_2e.

The carbon savings from generation depend on the carbon intensity of the electricity that you're replacing. Let's assume that your generation replaces the coal-fueled part of the country's energy mix. In other words, if you live in the US, let's say that rather than replacing typical grid electricity, which comes from a mix of gas, coal, nuclear, and renewables, the effect of your turbine is to reduce the use of coal power stations. In this case, the carbon saving is roughly 1.06 kg (2.3 lbs) per kWh, so you save over 30 tons per year and pay back the embodied carbon in a single year—a great start.

Rated power	Electricity generated over lifetime (MWh)	Embodied emissions (t CO_2e)	Carbon payback over 20-year lifetime (t CO_2e)	Payback period (days)
50 kW	2,260	59	(-) 1,317	312
100 kW	4,520	134	(-) 2,619	354
250 kW	11,300	148	(-) 6,734	157
500 kW	22,601	274	(-) 13,490	145
900 kW	40,681	289	(-) 24,486	85
2 MW	90,403	937	(-) 54,119	124
3 MW	135,605	1,046	(-) 81,538	92

Embodied emissions and carbon payback of macro wind turbines

With macro-renewables, things are even better. A recent paper analyzed the embodied emissions and carbon payback of onshore wind turbines ranging from 50 kW up to 3 MW in size (see table above). They included manufacturing, transport, and installation emissions in their scope, and found that a 100 kW wind turbine has an embodied footprint of 134 tons CO_2e, and a 3 MW wind turbine has 1,046 tons of embodied carbon.[9] That sounds like a lot, but the 100 kW turbine is expected to generate 213,000 kWh per year, and the 3 MW turbine over 6.5 million kWh (assuming they are placed in a sufficiently windy area). This would give a carbon payback of one year for the 100 kW turbine and just ninety-two days for the 3 MW turbine. But, as the table above shows, all large wind turbines pay back quickly, and generally speaking, the bigger the turbine, the better.

A new-build house

32 tons CO_2e 3-bedroom UK townhouse, bricks and mortar
53 tons CO_2e 4-bedroom UK detached, bricks and mortar
72 tons CO_2e 4-bedroom US detached, bricks and mortar

> We need to do better with house building

The figures above are based on my analysis of houses built a few years ago by one of the UK's big developers. They were designed to pass through the building regulations of the day, which can only be described as inadequate for the low-carbon world. In the US, a four-bedroom house is much larger in square footage than in the UK, and with that comes a higher construction footprint.

About a quarter of total footprint comes down to "site works," which includes landscaping, building roads in a new development, and getting the electricity and other utilities to the house. The foundations add about 5 percent and the bricks and block construction about 15 percent. Then there's 3 percent for the roof and 4 percent for doors and windows. At this stage we have something that looks like a house from the outside, but we are only halfway through accounting for the eventual carbon footprint of the finished item. Plastering is a huge deal, coming in at nearly a quarter of the total footprint, because plaster is chemically similar to cement and has the same emissions issues (see *A bag of cement (25 kg/55 lbs)*, p. 115). Stairs, floors, and other timber add 7 percent; plumbing and electrical add 10 percent; and kitchen and major appliances another 4 percent. Insulation, which radically cuts the carbon footprint of living in the house, adds just 1 percent.

For houses that are built with the climate emergency in mind, we should probably be adding between 20 and 40 percent to the footprint for solar panels across at least

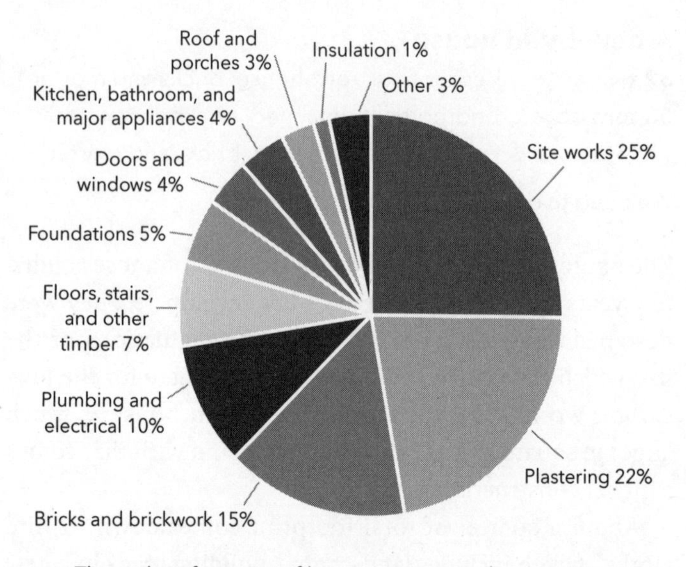

The carbon footprint of houses on a typical new estate

one aspect of the roof (say, 8 kW peak capacity for a four-bedroom house) for heat pumps, a battery, and, of course, significantly more insulation. But, against this, there is also scope for big reductions in the footprint by using different materials such as hempcrete, timber for the main structure, and even straw (I would love to build a straw house, even if only for the fun of it).

When I wrote the first edition of *Bananas* in 2010, the carbon footprint of living in a house typically overtook the footprint of building it after just a few years. Going forward, that shouldn't be the case. While the footprint of construction can come down, houses can and should generally be built to be powered entirely by electricity, and even to generate *more* electricity than they use.

In energy performance terms, the UK has an old and leaky housing stock, and this won't change very fast—each year we only knock down about one out of every 3,000 houses, because it's such an expensive and wasteful thing to have to do, while we add about 1 percent to the total through new construction. Although the average US house is significantly younger and less leaky than its UK equivalent, it is still not as efficient as it needs to be. It is a great shame to build a new house without the best energy credentials, because once it is built, we are stuck with it and retrofitting is so much harder than getting it right in the first place.

However, since most of us live in older houses, let's have a look at what we should do with them. A few years ago, my company looked at the carbon implications of various options for a traditional cottage in Dumfries, Scotland: leave it as it is, refurbish it, or knock it down and build a new dwelling. We looked at the climate change impact over a hundred-year period, taking into account the embodied emissions in the construction and maintenance, as well as living in the building.

The worst option by far was to do nothing and leave the old house leaking energy like a sieve. Knocking down and rebuilding worked out at about 80 tons CO_2e (whether you followed Scottish building regulations or the much more stringent Code for Sustainable Homes Level 5 that demands carbon neutrality). For these new-build options, the upfront emissions from construction work were paid back by savings from better energy efficiency in fifteen to twenty years. However, the winning option was to refurbish the old house. This involved a carbon investment of just 8 tons CO_2e, and, once cost was taken into account, refurbishment became dramatically the most practical and attractive option, too.

If this study is representative, as I suspect it is, then investing in improvements to existing old homes is a dramatically more cost-effective approach than knocking down and starting again. It is also a huge potential source of green jobs.

A car crash

0 CO_2e a tiny bump that you can live with
5 tons CO_2e a write-off on an empty road
40 tons CO_2e a double write-off on a busy highway

> A car crash is a very high carbon expense

If you drive, try not to crash. Health and welfare aside, it will give your carbon footprint an almighty dent. If you were to write off your car—with a value of, say, $20,000—on an empty road, without damaging anyone or anything else, that's your year's 5-ton carbon budget, based on embodied emissions of the car's manufacture of 267 g CO_2e per $1 of value (see A *new car*, p. 157).

If you're involved in an accident, things can escalate fast. Let's say you write off two mid-sized cars, each worth $33,750 (that's 18 tons CO_2e), and cause a ten-mile traffic jam across three lanes of a highway for two hours. If the queue has frequent stops and starts, the 6,000 or so cars involved will be unable to turn their engines off and will each emit perhaps 4 kg (8.8 lbs) CO_2e more from their exhaust pipes. That adds up to a further 24 tons of emissions.

My sums have not taken account of the footprint of a whole string of other consequences of the crash: the extra burden on the emergency and health services, congestion on surrounding roads, wear and tear on the cars, to name but a few. If complex surgery were involved for one or more injured drivers, that could boost the total footprint significantly (see *An operation*, p. 146). I should also mention writing off a big SUV would be roughly double the footprint of writing off an average-sized car. If the crash involves another vehicle, you'd also be more likely to seriously injure someone (see *New York City to Niagara Falls and back*, p. 124), racking up further emissions from the hospital treatment and surgery that follows.

It is also interesting to look at the human impact of the crash. Imagine, for the sake of argument, that one fifty-year-old person dies but everyone else is more or less fine. We could say that the human impact has been the loss of thirty person-years of life (plus a massive impact on the lives of friends, families, and colleagues). As for the broader impact, if the 6,000 cars each had a typical 1.6 occupants, about 20,000 person-hours will be spent in the living hell of traffic gridlock. That's a total of three person-years of life lost. Finally, there are the victims of climate change. The extent to which people around the world are affected by the release of 40 tons CO_2e is impossible to quantify but must equate to a few more years lost.

Having a child

300 tons CO_2e baby born in 2021 into a family with a US-typical footprint, which is then cut by 10 percent annually

1,260 tons CO_2e same baby, same family, but without the 10 percent annual decrease in footprint

5,000+ tons CO_2e baby who grows up to be both wealthy and carbon-careless

> The more children, the higher the carbon, but by how much depends on how they live

Unless you ever light a bushfire (see p. 185), the decision to reproduce is probably the biggest carbon choice you will ever make. The more of us there are, the greater the pressure on the world's resources. I'm not saying you shouldn't have children. And, if you are someone who believes that God has told you to go and have ten of them, I am not even saying that you are wrong about that. All I'm saying is that one of the consequences will be thousands of tons of carbon emissions.

In my 300 tons CO_2e low-emissions scenario, our child is born into a household that has a US-average carbon footprint per person. However, the household cuts its carbon in line with what the government really ought to promote and prescribe as "following the science"—achieving a footprint that will give us a 66 percent chance of keeping climate change to 1.5°C.

This is, of course, a tricky concept to model, as nobody knows for sure what that overall global CO_2 budget should be. But there are various estimates, and for these sums I've gone with a figure of 350 gigatons CO_2e (a gigaton is 1,000 million tons), based on a spread of estimates and giving a higher weighting to some of the more optimistic ones because I think they are based on more recent analysis.[10] To achieve this, it will mean our child cutting its US CO_2 footprint by just over 10 percent every year throughout its lifetime (which I have assumed to be ninety years, since the world is a healthy place). I have applied a less aggressive approach to the other greenhouse gases, cutting them by just 3 percent per year.

For the middle scenario, I've assumed that our child lives its life as a carbon-average US citizen, but the US fails to cut emissions. This is a dreadful scenario for the world, and with unknown consequences, but it seems realistic to give this person a life expectancy of sixty rather than today's eighty years.

At an even higher end of our baby budget are children who, even if you have done your best to encourage sustainability values, go on to lead high-carbon lives. Although they are quite well-off, their brains don't work properly, so they manage to go through life without caring about climate change despite its obvious effects, or perhaps lack empathic connection with the rest of humanity and don't understand that carbon responsibility applies to them as well as anyone else. They perhaps drive an SUV, assuming these aren't banned, take regular long-haul flights, and don't pay any heed to carbon considerations in their diet or any other part of their life.

All my scenarios assume that the person lives in the developed world (the numbers would be much lower in developing countries). And, for simplicity's sake, I have not taken into account the footprint of their own offspring…

Millions of tons

A volcano

1 million tons CO_2e Mount Etna in a quiet year
5 million tons CO_2e Holuhraun eruption, Iceland, 2014
42 million tons CO_2e Mount Pinatubo, Philippines, 1991[1]

> Volcanoes are high carbon, but also have a cooling effect on the planet

Has anyone ever tried to convince you of the belief that human-generated carbon emissions are dwarfed by those from volcanoes? It's a common myth, and it's nonsense. The world's volcanic activity, from 400 or so active volcanoes plus volcanic lakes and rifts, produces a total of 280 to 360 million tons CO_2 per year.[2] This is less than 1 percent of the annual emissions from human activities.

Nonetheless, as the figures above show, each active volcano does have a massive footprint, with a major eruption causing tens of millions of tons CO_2e. But these numbers are misleading, because (alongside their warming effect) volcanic emissions also cause cooling. The ash and sulfur dioxide that they throw up into the stratosphere reflect sunlight away from the Earth. Overall, the Mount Pinatubo eruption of 1991 is thought to have resulted in a net planetary cooling of 0.5°C the following year.[3] Over time the cooling effect fades faster than the greenhouse effect of the carbon, so the question as to whether the warming effect or the cooling effect is greater is not clear cut.

The World Cup (soccer)

2.2 million tons CO_2e 2018 Russia World Cup[4]

2.7 million tons CO_2e 2014 Brazil World Cup

2.8 million tons CO_2e 2010 South Africa World Cup

> The 2018 tournament produced emissions equivalent to a half pint of beer for every man, woman, and child on the planet

The footprint figures here, as you'll see from our pie chart, include players and their entourages traveling around, the construction of the sites, energy used at the stadiums, accommodation, and fans traveling. For the 2018 tournament in Russia, an estimated 3 million spectators saw matches live, at an apparently massive carbon cost of 700 kg

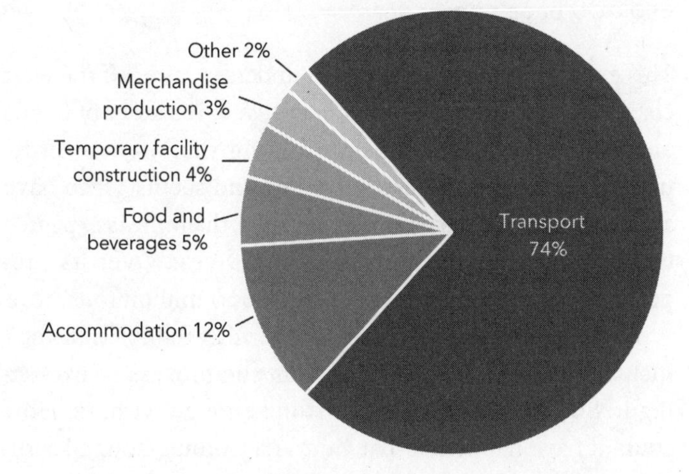

The carbon footprint of the 2018 World Cup in Russia

(1,540 lbs) per spectator.[5] However, the sixty-four matches were viewed on television and online by 3.6 billion people worldwide (with a total of 34.7 billion viewer-hours)[6] and that comes to only 62 g per hour of entertainment per spectator, not accounting for the viewers' footprints—their TVs, computers, and pints of beer or cups of tea—while they watched the sport.

By comparison, a UK Premiership soccer match comes out lower carbon, because there is far less international travel, and again a large TV audience (90 percent of it outside of England). Even better—healthier and arguably more fun—is a kick-around in your local park or street. This is virtually emissions-free.

A new coal mine

8.4 million tons CO_2e Woodhouse Colliery, Cumbria, UK
(annual emissions for 50 years)

> One small coal mine can equal the annual carbon footprint of 650,000 people–each year

This entry is based on a proposed undersea mine off the west coast of Cumbria, UK, near where I live. It is a story of bogus analysis and terrible decision-making that went weirdly under the radar for a very long time and seems set to have appalling consequences. If it goes ahead, the mine is expected to extract 2.3 million tons of coal per year. Over its proposed fifty-year life, this will result in 420 million tons CO_2e.

And this is just the footprint for burning the coal. I haven't included the lesser emissions from the process of extracting it. But the numbers are startling enough without. How could it have happened that Cumbria County Council actually approved planning permission and in November 2019 (despite having committed to zero carbon by 2050) the government declined to revisit the decision? The answers take a little bit of unpacking.

The planning argument for the mine ran as follows. Although the UK has committed to phase out coal for heating and electricity, this mine is intended to extract special "metallurgical coal" used to make coke for steelmaking. Global steel production is (the proposal claims) set to rise, and metallurgical coal demand will have to go up in step with this. Woodhouse will supposedly not lead to an increase in global coal extraction, because other coal mines will cut production by an equal amount and the result will be reduced shipping emissions compared to importing coke from Australia. Finally, the application suggested that the mine might provide 500 jobs in a part of the world that badly needs them.

None of these arguments has merit. Europe's steel producers almost all have plans to decarbonize 100 percent by 2050, which will leave this mine stranded. Technologies for producing steel without fossil fuel are just about to come fully on stream (see A *ton of steel*, p. 148). The claim that other mines around the world would reduce their output to compensate is made without evidence and runs entirely against economic theory. The savings in shipping emissions would be dwarfed many times over by emissions from burning the coal. Which leaves only the question of jobs. Cumbria does indeed need more high-quality jobs. But what it really needs is thousands of green energy jobs, not a few hundred coal-mining jobs that come at a carbon cost of 16,500 *tons* CO_2e *per employee per year*.

As I write this, the mine is still set to go ahead, and in doing so it will knock a large hole in the UK's climate credibility. For more, see the Green Alliance report.[7]

Cryptocurrencies

46 million tons CO_2e Bitcoin in 2019

68 million tons CO_2e all cryptocurrencies in 2019[8]

> In just a decade, cryptocurrencies have eaten up 0.12 percent of the world's carbon footprint[9]

Cryptocurrencies, which launched with Bitcoin in 2009, have to be one of the most fundamentally pointless ways of using energy. Along with space tourism, they surely provide one of the best illustrations of humanity's seemingly limitless appetite for energy.

Digital currencies are made up of virtual coins that are created or "mined" by computers solving puzzles. In order for these currencies to have value, the coins must be scarce, and to achieve that the puzzles must be hard. But, as the world's computer power goes up, and as the number of computers devoted to mining rises, so must the difficulty of the puzzles. The result is like an arms race between the complex problems and the stacks of purpose-built mining machines, whirring away day and night. Because miners compete with each other to solve the problem fastest, more and more sophisticated farms of servers are built. This leads to miners burning through a huge amount of electricity—currently a whopping 0.3 percent of global electricity use for Bitcoin alone and 0.5 percent for all cryptocurrencies.[10]

To make matters worse, most cryptocurrencies, including Bitcoin, use a system called "blockchain," which records transactions digitally in so many places around the world that they cannot be forged. This huge level of duplication (known as a "distributed ledger") adds even more to the energy use and the carbon footprint. Not only is the carbon footprint of Bitcoin as a whole shooting upward, but so too is the carbon footprint of each individual coin.[11]

In my calculation of an annual footprint of 68 million tons CO_2e for all cryptocurrencies, I've assumed a global average for electricity used. This may be underestimating the footprint if mining is often done in countries with coal generation (as some suggest), though one Bitcoin investment fund[12] argues that its electricity is largely from renewables. I haven't included the embodied carbon in the manufacture of the machines, as, unlike for personal IT, I suspect this will be a minor consideration compared to the power they use.

No one knows whether cryptocurrencies will continue to rise or die back. Some are worried that, if the growth trend continues, Bitcoin alone could push the world over 2.0°C warming within the next twenty years.[13]

Carbon aside, debate rages over the role of cryptocurrencies in the world. Their anonymity makes them great for money laundering and very useful on the darknet. Overall, the world is surely better off without them.

The Cloud and the world's data centers

160 million tons CO_2e in 2020[14]

> Data centers use about 1 percent of global electricity and 0.25 percent of its footprint[15]

Data centers are buildings packed top to bottom with computers that make our information age possible. They store the Cloud with all its web pages, databases, applications, photos, video content, and other downloads. As you'd expect, they use lots of electricity (both for powering the machines they contain and for keeping them cool with air conditioning) and, as people consume ever more digital content, their already considerable carbon footprint is continuing to rise.

Ten years ago, for the first version of this book, I put the estimate for data centers at 130 million tons and wrote that, on current growth trends, this was set to at least double over the coming decade. Actually, the footprint has grown by less than a quarter, since growth in usage has only slightly outpaced efficiency improvements and uptake of renewable electricity. My guess is that, in the absence of a global carbon constraint, the trend will continue.[16]

Meanwhile, it's interesting to note that the UK's footprint accounted for by print-based publishing stands at about 1 percent. A direct comparison is complex, but you would expect digital information replacing print to result in a reduced carbon impact. That may not be the case due to the so-called rebound effect—the idea that when something (in this case, data) becomes cheaper and more carbon efficient to do, we end up simply doing more of it.[17] With data, it may be that the reverse is true. Not only is global data growing incredibly fast but so is our expectation that we can interrogate it at a moment's notice. We take, send, and store multiple copies of high-resolution photos and videos with barely a thought, and stream video content whenever we feel the smallest urge. And, where we might have expected to queue at busy times in a bookstore to inquire about the contents of its shelves, Amazon now has to meet our expectation through its data center capacity, so that even in peak times we can search the whole of the world's published materials in an instant.

Of course, if we go for digital information without ditching the paper, downloading stuff simply to print it out, we end up with the worst of all carbon worlds. The digital world has dented but not totally replaced the paper world.

Data centers are just one component of the footprint of the digital world (see *The world's* IT, p. 187).

The USA (and other countries)

352 million tons CO_2 UK CO_2 emissions (within its borders)
435 million tons CO_2e UK overall greenhouse gas emissions (within its borders)
840 million tons CO_2e UK greenhouse gas footprint (including imports, etc.)[18]
6 billion tons CO_2e US greenhouse gas footprint (EPA estimate)[19]
7.6 billion tons CO_2e US greenhouse gas footprint if you include all imports, trade, and account for truncation error

> The US emits up to 7.6 billion tons CO_2e a year, making it the second largest emitter in the world

You can get very different numbers for a country's emissions, depending on what you count and how you do the counting. For the figures above, I've used UK government data for emissions of CO_2 and other greenhouse gases within its borders. But these official statistics don't include the carbon footprint of goods we import (less the smaller amount that we export), nor aviation and shipping. My total greenhouse gas footprint estimate also includes a markup factor for aviation (see p. 152), to take account of high-altitude emissions. When combining all the gases into one metric (CO_2e), I've chosen a hundred-year time period. However, if we are concerned about a shorter timescale, as we should be, most of the other gases, which have shorter lives than CO_2, become more important; their effect roughly doubles if you look at their impact over the next fifty years. For lower numbers for the US, I have used the EPA's reporting for greenhouse gas emissions and sinks for 2018. However, this is unlikely to account for all trades and imports and will have cut off some form of truncation error in the supply chain emissions, so I have applied a markup factor to get the high-end number.

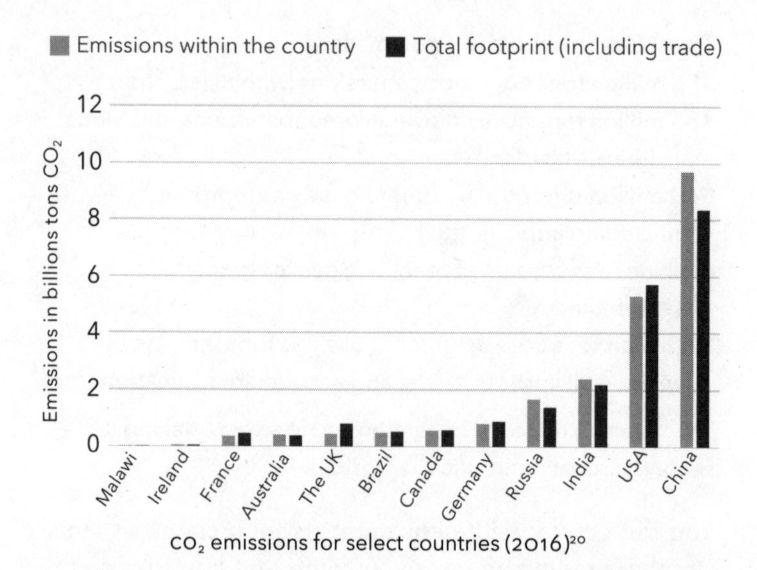

CO_2 emissions for select countries (2016)[20]

To compare different countries—their direct emissions and their total carbon footprints—I've only been able to get ahold of data for CO_2 emissions for 2016. But the chart above tells an interesting story, both in terms of who the big emitters are and which countries are net importers (those whose footprints are bigger than the emissions that take place within the country itself). The UK and US stand out as a big importers, while China and Russia are notable exporters.

Of course, the countries listed have very different populations. To get a perspective on the relative carbon extravagance of different nations, we really need to look at the consumption footprint per person (see chart opposite). The highest footprints are those of the US, Canada, and Australia, with western European nations in the tier below and China below that—but rising. After that, there is a big jump down to Brazil and India and, finally, Malawi is almost invisible.

Yet another way of looking at a country's footprint is in terms of emissions per unit of GDP (gross domestic product), as shown in the chart overleaf. This is a measure of carbon efficiency or carbon intensity—a nation's footprint relative to its economic activity. Countries with inefficient factories, and which get their electricity from dirty coal-fired power stations, rate worst on this scale. Hot countries can sometimes achieve a better rating, because they don't have to spend so much on keeping warm (provided that people aren't rich enough to afford air conditioning).

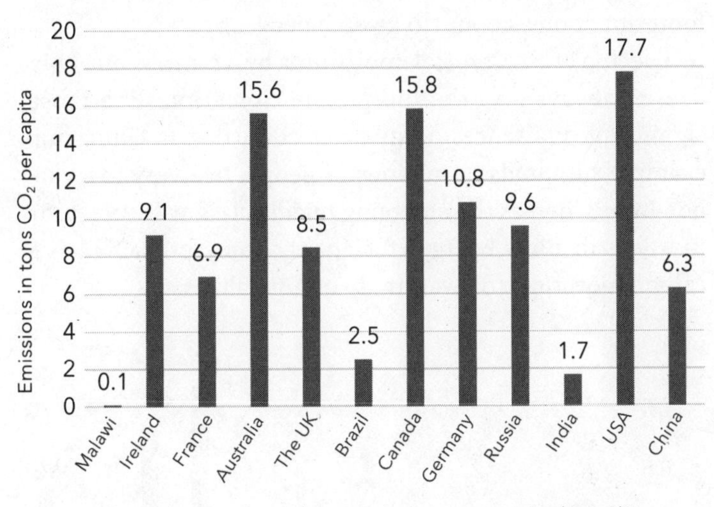

Consumption-based CO_2 emissions per capita (2016)[21]

Through this GDP lens, Russia comes out worst because of its coal-fired power stations, inefficient factories, and cold climate, with coal-dependent India and then China following behind. Western Europe has relatively efficient factories and cleaner electricity, so countries in this region come out well, especially nuclear-powered France. The US, Australia, and Canada come out somewhere in the middle of the carbon efficiency stakes. The UK has improved considerably in the ten years since the first edition of this book and has just about managed to decouple economic growth from emissions. The world as a whole still shows little sign of achieving this. There are many good reads and talks on this subject (see endnote).[22]

Ultimately, there's no avoiding the fact that a country's emissions are strongly linked to its wealth. It's hard to be rich and have a low-carbon footprint (see *Spending* $1, p. 79). Malawi is just one example of a country whose poverty ensures a low footprint. In 2016, its 17 million people had a footprint of only about 110 kg each per year.

I've looked at typical footprints by country, but this doesn't always give the full picture. Sometimes the most significant differences occur *within* countries. In China, for example, hundreds of millions of people live very low carbon lives, whereas the emerging middle class, with Western lifestyles in a less energy-efficient economy, probably have carbon footprints to dwarf those of Australians.

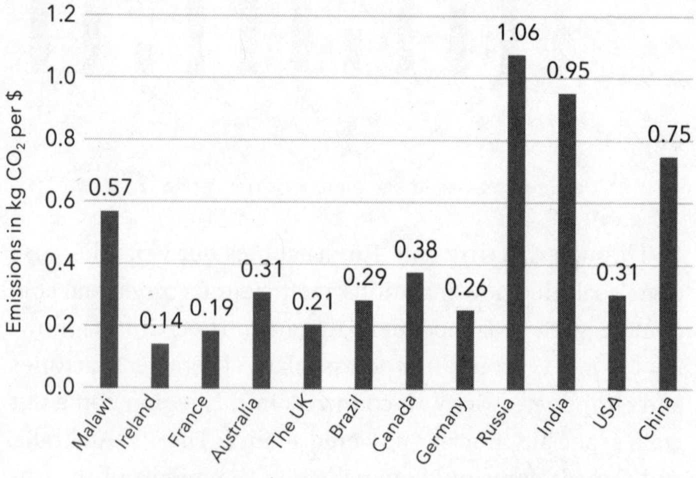

CO_2 emissions by country by GDP (2016)[23]

Billions of tons

Wildfires

259 million tons CO_2e US wildfires in 2020[1]
923 million tons CO_2e Australian bushfires in 2019
8.6 billion tons CO_2e global wildfires in 2019

> Bushfires in Australia accounted for 2 percent of the world's carbon footprint in 2019

Australia has always had bushfires, affecting its grassland, savannah, and shrubland. Average annual emissions from these types of fires between 1997 and 2018 were 448 million tons CO_2e per year, and that was roughly unchanged in 2019. What was different was that the country's temperate forests caught fire as well, adding half a billion tons CO_2e, turning the skies orange 3,000 kilometers away in New Zealand, and destroying 17 million hectares (42 million acres) of forest along with most of its wildlife. These *new* fires alone added 1 percent to the world's entire carbon footprint.

If you were to start a wildfire deliberately, that one strike of a match would make your footprint thousands of times greater than most people build up over their lifetimes. However, arson isn't the main driver for the fires in Australia. That would almost be reassuring. No, the prevalence of fires is almost certainly a symptom of climate change and just a taste

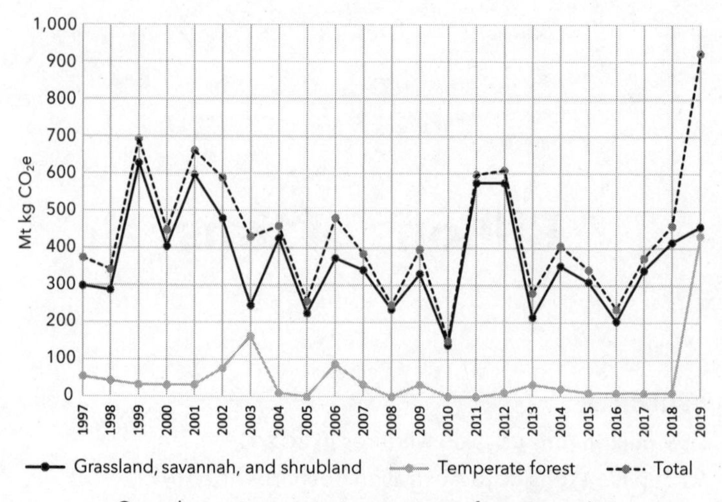

Greenhouse gas emissions per year from Australia's bushfires between 1997 and 2019, by type of forest

of what is to come. Wildfires are an early example of a positive feedback loop in the climate system: emissions cause warming, which causes fires, which release yet more emissions. We call it a climate emergency because it is one. And, even if we take rapid global action now, there is nothing at all to stop symptoms like this getting more extreme for at least a few decades until the world reaches net zero emissions.

Australia, of course, is not the only place subject to wildfires; there are huge problems each year in Indonesia, for example, and areas prone to wildfires will increase as our climate warms. One particular source of alarm was that in 2019 the Amazon saw a 75 percent increase in fires,[2] and a recent paper in *Nature* predicted that this huge carbon sink could soon turn into a net carbon source,[3] especially if climate change–induced fires continue to be supplemented by deliberate deforestation fires (see *Deforestation*, p. 191). Globally, in 2019, emissions from wildfires came to 8.6 billion tons CO_2e, a mix of what we might call natural fires and human- or climate change–induced ones.

We saw a similar picture in the US in 2020, with the wildfires raging through California set to be the worst on record. About 260 million tons CO_2e were emitted from all fires in the US in 2020, or nearly four times the 1997–2019 average of 70 million tons. Climate change usually affects the poor disproportionately. However, the wildfires in California, including prime real estate around Los Angeles, serves as a reminder that no one is immune from global warming.

The world's IT

1.4 billion tons CO_2e total including all user devices, data centers, networks, and TV

> IT now accounts for 2.5 percent of global emissions

We live in a digital age, so you might expect the world's IT to account for a larger share of our emissions. But the numbers are still pretty staggering. The 1.4 billion tons CO_2e total is split more or less equally into three parts: the first includes phones, computers, consoles, and tablets; the second comes from TV; the rest is data centers, networks, and, shockingly, 5 percent is from cryptocurrencies.

These estimates come from a careful picking through of the relatively small amount of research on the carbon footprint of IT.[4] It is fiendishly difficult to come up with accurate numbers for a sector that includes everything from TVs and their broadcasting networks to all computers, phones, tablets, and game consoles, the data centers that store and analyze the world's information, and the networks that transmit it. I've included the electricity used (using a global average carbon intensity) as well as the carbon footprint of manufacture and disposal of all the kit. I haven't included, because I wouldn't know where to start, what is often termed the Internet of Things—that is, the computing capacity that is increasingly built into everything you can think of from your car to your washing machine. For data centers and networks, the energy they use is the dominant factor, whereas for user devices, most of the carbon lies in their manufacture.[5]

A debate has raged about whether IT's footprint is going to keep on going up, stabilize, or come down, and also as to whether IT overall adds to global emissions, or whether it reduces the world's carbon footprint by enabling everything we do to become more efficient. One way of looking at things is through the lens of the Jevons paradox.[6] This is the idea that efficiency gains in an industry tend to bring about increases in total environmental burdens because they make possible a greater increase in usage than can be mitigated by the efficiency gain. So, the question is whether or not the Jevons paradox applies to IT itself and to IT's role in the world.

Looking back over the last seventy years, it is clear that four things have gone hand in hand:

- IT has become more efficient by many thousands of times.
- IT's footprint has gone up by a factor of many thousands.
- IT has enabled enormous efficiency improvements.
- The world's carbon footprint has continued to go up and up.

A couple of storylines about IT need to be put into context. The first is the idea that video conferencing cuts carbon by enabling people not to fly. This same enablement argument is applied to dozens of efficiency improvements. Unfortunately, the evidence is against it because (over the past few years) the carbon footprint of flying has been going up rather than down and, while we might avoid a flight by having a Teams, Zoom, or Skype call, we might also have a few of those calls with someone and then decide to meet up. (Whether the new ways of global communication during and after COVID-19 will change this pattern remains to be seen.)

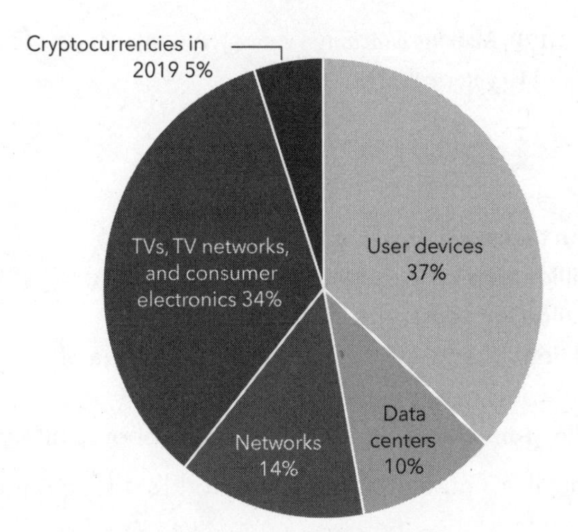

The world's IT footprint (1.4 billion tons CO_2e), 2020

The second storyline is that IT will stop growing as we become saturated with it—when everyone has as many personal devices as they could possibly want or need. In some parts of the world, we are perhaps beginning to reach that stage with many devices. However, just as saturation effects arguably begin to show up in these more conventional forms of computing, they coincide with an explosion in newer strands such as artificial intelligence, cryptocurrencies, and the Internet of Things.

Looking to the future, the IT industry is set to carry on growing and its carbon footprint will only come down if the industry makes that happen. Some of the giant tech companies are starting to make some serious pledges in the right direction, but whether self-regulation can be enough remains to be seen. And while IT doesn't automatically cut carbon in the rest of the economy (and perhaps does the opposite by default), it can make things a lot easier for us should we finally decide to leave the fuel in the ground. (See also *Using a smartphone*, p. 126, *The Cloud and the world's data centers*, p. 179, *An hour watching* TV, p. 47, A *computer (and using it)*, p. 140, and *Cryptocurrencies*, p. 178.)

A war

9 million tons CO_2e UK military boot print in 2019
200 million tons CO_2e US military operations in 2019
400 million tons CO_2e Iraq War, 2003–19
690 million tons CO_2e "limited" nuclear exchange of 50 15-kiloton warheads[7]
3.3 billion tons CO_2e annual global military boot print[8]

> Note that final figure–it's double the world's IT footprint

The direct human costs of wars are so great that it might seem flippant to think about their climate change costs. But war unfortunately plays a big role in global society, so it needs to be considered. Moreover, it's worth bearing in mind that the emissions of a war could ultimately have serious climate change impacts.

The financial cost of the US military operation in Iraq between 2003 and 2019 has been estimated at $2 trillion. This includes spending by the Pentagon, but also the State Department, as well as medical care and disability compensation for veterans and interest on borrowing.[9] We can use this spend figure to give a crude estimate of the footprint of the US operation—340 million tons CO_2e. This excludes the actual emissions from combat itself and its impacts. You could add a few percent to both numbers to include coalition forces and perhaps another 1 percent for the much more poorly resourced insurgency. That's how I get to an overall estimate of 400 million tons.

With a total of $730 billion in 2019, the US spends more on defense than the ten next-highest military spenders combined. That's 36 percent of the world's total military spend.[10] Its footprint of 200 million tons CO_2e makes the US Department of Defense the single largest institutional producer of greenhouse gases in the world.[11]

In 2019, the UK spent "only" $49 billion, with a footprint of "just" 9 million tons CO_2e in 2019. One estimate of the total impact of war, when all the indirect impacts[12] are taken into account, is 3.3 billion tons CO_2e, or 6 percent of global emissions.[13]

I wrote in the first edition of *Bananas* about a study that estimated that the soot from a small regional nuclear conflict with fifty warheads could cause a net cooling effect over the first few years.[14] However, new studies estimate that there could actually be a huge net cooling effect of 2–5°C globally in the case of a regional conflict, and up to 10°C of cooling from a US-Russia nuclear war.[15] Climate change, of course, might well be making nuclear war more likely.[16]

Looked at in the starkest and simplest possible terms, even with a non-nuclear war there would be carbon savings from people ceasing to exist, which might make up for the direct emissions from the war in just a few years. In other words, mass annihilation turns out to be an effective way of curbing emissions, though, of course, it also defeats the object.

Deforestation

(-) 500 tons CO_2e UK reforestation per hectare
((-) 1,250 tons per acre)
1,000–2,000 tons CO_2e deforestation per hectare
(2,500–5,000 tons per acre)
5–10 billion tons CO_2e global deforestation per year

> Each deforested hectare is equivalent to driving a car fifty to a hundred times around the world

A hectare is roughly one-and-a-half soccer pitches. To be exact, it is 100 m × 100 m (330 ft × 330 ft), so there are a hundred hectares in a square kilometer and about 260 in a square mile. At present there are 4 billion hectares of forest on our planet, or about 25 percent of the world's land area,[17] and they store more than a trillion tons of carbon.[18] This makes them one of nature's most important carbon sinks, along with peatlands and wetlands.

More than half of the world's forests are found in only five countries: the Russian Federation, Brazil, Canada, the US, and China. And, as we all know, deforestation has been proceeding at an alarming pace. Since 1990, it is estimated that 420 million hectares (more than 10 percent of the global total) have been lost. Currently, the rate of loss is estimated at around 5 million hectares annually, much of it primary forest in Brazil and Indonesia.[19] That's a bit more than the size of Switzerland every year.

About 27 percent of all tree loss is to make way for commodities, such as palm oil, soy (for animal feed), beef, minerals, oil, and gas. A further 26 percent is cut within managed forests and tree plantations expected to regrow after harvest. Another 24 percent is cleared and burned for short-term cultivation of subsistence crops, mostly in tropical regions. Natural and climate change–induced wildfires make up a further 23 percent and 1 percent is to make way for creeping urbanization.

On the positive side, since the first edition of *Bananas*, the United Nations REDD (Reducing Emissions from Deforestation and Forest Degradation) Programme has been set up and has made big inroads, partnering with sixty-five countries in South America, Africa, and South Asia to support them in protecting their forests and achieving climate goals. On the negative side, deforestation has shot up in Brazil since the election of Jair Bolsonaro as president.

In reforestation projects, the carbon per hectare varies considerably, depending on the age, species, spacing, and soil conditions. So, the sums aren't easy. A five-year-old UK broadleaf plantation, which will store anywhere between 5 and 23 tons CO_2e per hectare (depending on the soil and the spacing), if left to grow for 200 years, will capture a further 600–1,350 tons CO_2e.[20] A capture figure of 500 tons CO_2e per hectare seems typical for UK reforestation projects, such as the Woodland Carbon Guarantee, which welcomes volunteers. In the US, check out One Tree Planted, a not-for-profit environmental charity that works in partnership with the US Forest Service.[21]

Black carbon

9 billion tons CO_2e globally per year

> Black carbon is hardly ever mentioned, but adds 16 percent to global emissions

Black carbon is a fine particulate matter that we get from burning carbon-rich materials like wood and fossil fuels—soot, in other words. It is caused by incomplete combustion. Forty-two percent comes from outdoor fires of one kind or another, and a quarter comes from the burning of wood, coal, dung, peat, and other organic stuff in homes. A further quarter comes from transport (mainly diesel) and about 10 percent from coal-fired power stations.[22]

Black carbon has a huge, but often overlooked, footprint. Some scientists consider it could be second only to carbon dioxide in terms of its climate effect.[23] But it is not included in most carbon footprints, and it is understudied, as there is no agreed definition or measurement of its emissions and impacts. For this reason, the concept of a "black carbon footprint" has been proposed.[24]

Black carbon warms the world in two main ways. Up in the atmosphere, it contributes to the greenhouse effect. Down on the ground, it turns snow and ice murky and in so doing makes it absorb more of the sun's heat. It is thought to be a major contributor to the global reduction in ice cover, especially in the northern hemisphere.[25]

The figure of 9 billion tons CO_2e is based on 9.7 million tons of black carbon being released in 2014[26] and a global warming potential of 910 over a hundred years (much higher than carbon dioxide's global warming potential of one).[27] A small amount of black carbon can have a hugely disproportionate effect on the climate.

The good news about black carbon is that it lasts only a few days or weeks in the atmosphere. In other words, if we can reduce the amount we create, the benefit will be instant, not just in terms of climate but also in the quality of the air that we breathe. Easy wins can be made by using particulate filters on diesel engines and swapping inefficient open fires (wood or coal) for super-efficient stoves.

The world's annual emissions

38 billion tons CO_2 annual CO_2 emissions, excluding deforestation (2018)

56 billion tons CO_2e annual greenhouse gas emissions caused by humans (2018)[28]

> Amazingly, this figure is still going up, as if humans had never noticed climate change[29]

The IPCC (Intergovernmental Panel on Climate Change) was established in 1988 by the World Meteorological Organization and the UN with the objective to "stabilize greenhouse gas concentrations in the atmosphere at a level that would prevent dangerous anthropogenic (human-induced)

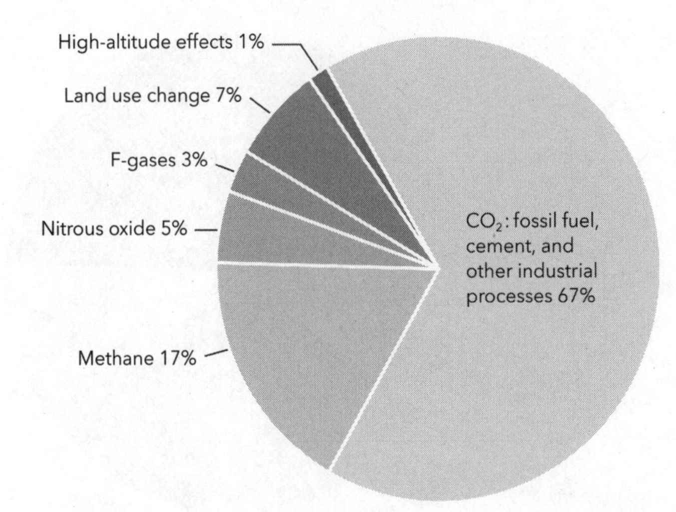

The world's annual footprint of 56 billion tons CO_2e (2018)

interference with the climate system." At the time, the world's greenhouse gas emissions were around 25 billion tons a year. When this book was first published, in 2010, emissions had reached 50 billion tons CO_2e. A decade along the line we have reached 56 billion tons CO_2e.

The 56 billion tons (for 2018) is the figure you get if you roll all greenhouse gases into the metric of CO_2e, based on the relative effect that each of the gases has on the global climate over a hundred-year period. If you are interested only in the very long-term climate impacts, many hundreds of years into the future, then only the CO_2 matters and it turns out that we can think in terms of a total all-time carbon budget for keeping to within 1.5°C of temperature rise. At current rates, we will have exhausted that budget by about 2030. We need to get to net zero—fast.

Many of the other greenhouse gases are more intense than CO_2 but don't last for that long in the atmosphere. They increase the *speed* with which the planet warms up, more than they affect the eventual temperature we might reach.

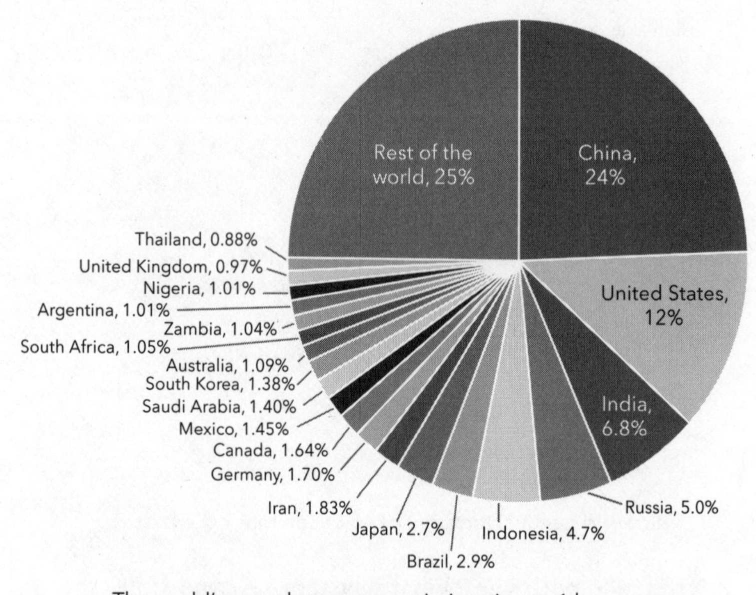

The world's greenhouse gas emissions in 2016 by country, showing emissions produced within each country's borders

So, if you are interested in the climate impacts that we might be experiencing in just fifty years' time instead of a hundred, then you need to roughly double the impact of the methane and the fluorinated gases (F-gases). If you are interested in where we might be around the year 2050, as well you might be, the methane is actually two-and-a-half times more important than it looks in the hundred-year analysis.

I'm going to stick with one hundred years, as that's the convention, even if it is debatable whether it is the most relevant. On this timeframe, CO_2 is about two-thirds of the story (before you include deforestation). Of this, coal is about 39 percent, oil 34 percent, gas 21 percent, and emissions from the chemical reaction in the cement-making process about 5 percent. Deforestation makes the CO_2 total up to almost three-quarters of our total footprint.

Of the methane, about a third comes from fossil fuels, a quarter from cows and sheep (enteric emissions), a sixth from waste disposal, a tenth from rice cultivation (anaerobic digestion in flooded paddy fields), and manure about 3 percent. About half of all nitrous oxide emissions are from manure. The rest is mainly from burning fossil fuels, chemical processes, fires of every kind, and agriculture.

We have to ask why on earth our footprint is still going up despite all the climate talks, protests, targets, and actions to cut carbon? The answer is that the global system is operating with self-correcting mechanisms so that, when one bit of the world cuts its carbon, it only encourages and enables other bits to increase their carbon an equal amount. That's a pretty stark reality, but I don't write it to be depressing. I write it because it tells us so much about what won't be enough and, more to the point, what we actually need to do to get on top of climate change.

It tells us that, at the end of the day, we need a global response. We need big system change. That has to include a global arrangement to leave the world's fossil fuel in the ground (see the following entry). As important as it is to ask, "How can I cut my carbon footprint?" we need to approach the question thinking "How can I push for the system change we need?"

Burning the world's fossil-fuel reserves

5.3 trillion tons CO_2e burning the "proven reserves" of conventional coal, oil, and gas

9.3 trillion tons CO_2e add tar sands and shale oil

45 trillion tons CO_2e burn all recoverable resources[30]

> This is the most important entry in the book: fossil-fuel reserves must stay in the ground

We all have to understand these two facts:

- Burning "just" the world's conventional proven reserves would cover ninety-five years of emissions at current global levels.
- This would trigger about 4°C of temperature rise *on top of* what we have already caused.

In other words, most of the world's fossil fuels cannot be utilized without causing catastrophic climate change. And yet they are on the balance sheets of the world's wealthiest countries and all of the major oil and gas companies. Most of these reserves are, in reality, valueless. Or, in fact, very much worse than that. They will cost the Earth. Climate change is real and it is happening now.

The "proven reserves" figure above is what fossil-fuel companies think they can extract and sell at a profit using today's technologies. Add the unconventional tar sands and shale oil, and the figure almost doubles. The total "recoverable resources," including everything that we know how to extract if we wanted to, doesn't bear thinking about. There is no getting around it: we *have* to leave nearly all these fossil-fuel reserves in the ground.

Unfortunately, the denial of this dazzlingly simple and important point by most fossil-fuel companies, most politicians, and most of *us* most of the time is the reason that we now have a flat-out climate emergency on our hands. Even now, most of the fossil-fuel companies are still investing in both exploration to find yet more fossil-fuel deposits and in new ways of extracting the stuff. If you have a pension plan, it is probably supporting some of the companies that are doing this.

Nobody knows in detail how things will play out. If we take very strong action now, there is a good chance that we

The world's fossil-fuel reserves

Unknown reserves ??

Other known and recoverable fossil fuel 36 trn tons CO_2e

Proven fossil-fuel reserves 9 trn tons CO_2e

Emitted 2 trn tons CO_2e

Budget remaining for 1.5°C rise, 0.4 trn tons CO_2e

The enormous CO_2 emissions bound up in the world's fossil fuel compared to our tiny remaining budget. The arrows show the ongoing process of discovering even more, digging it out of the ground, burning it, and squandering the last dregs of our budget.

can deal with the fallout of what we've already burned without it being too disastrous. But one thing is for sure: if we don't make huge changes to the way we live, we will definitely be in a great deal of trouble.

Negative emissions

> It's no longer enough to reduce emissions—we need to start removing CO_2

If the world gets to net zero carbon emissions by 2050, it looks as if we will stand a reasonably good chance of avoiding a climate disaster. But, to achieve that, it won't be enough just to cut emissions. We will need to actively remove CO_2 from the atmosphere. And we should start right now.

What follows, then, is a quick introduction to the main options for carbon removal (or offsets or negative emissions) available to us as individuals, companies, or countries. All come with one or more of the following problems: a fundamental limit to the amount we can deploy them; the creation of other environmental pressures; risks of going horribly wrong; and a dependence on technologies that are unproven at scale. They are definitively not *alternatives* to cutting our carbon emissions, which will always be better and almost always far cheaper than carbon removal.

We should certainly not fall for the idea that we can simply offset our carbon footprint, or, as is often offered by airlines, our flights. Such offsets are often marketed at absurdly low cost. In 2019, the global voluntary offset market surged to $300 million and claimed to offset a whopping 100 million tons CO_2. That's just $3 per ton of carbon saved.

To achieve these prices, the schemes look to enable others (generally in the developing world) to reduce their carbon footprint, through providing more efficient cooking stoves, for example, or funding a renewable energy project. These can be good things to support, but they aren't negative emissions or genuine offsets. They are projects that need to happen anyway, and very often the actions don't cause the savings they claim. Even schemes for avoiding deforestation aren't real offsets, though of course it's crucially important to support rainforest maintenance and reforesting.

"Net zero" targets have become popular for companies, countries, and individuals. The first thing to say about these is that, for both organizations and individuals, they need to be accounted for with both a carbon-cutting target *and* a carbon-removal target. Otherwise, there is a strong temptation to use offsets as an excuse to shirk our responsibility to cut the emissions in the first place. Only once all possible emissions reductions have taken place can it start becoming reasonable to talk about offsetting the remainder to achieve net zero. The next test for a legitimate offset is that it must actually remove carbon from the air. After that, you need to show that the removal wouldn't have happened without your funding (the so-called additionality test). Finally, you need to check for any other negative environmental or social consequences or risks.

If it comes to trying to offset any avoidable emissions, such as an unnecessary flight, to my mind there is a further and very tough test to apply. To stand a hope of a legitimate offset claim, you need to show that the carbon you remove is genuinely *additional* to everything that needed to happen anyway without your avoidable impact. This last test rules out all the finite nature-based solutions (tree planting, soil sequestration) that humanity already needs to maximize and, as we will see below, takes us into unproven technologies costing at present hundreds of dollars per ton.

In turning now to the negative emissions options, note that the figures I've used for the potential of each option are high-end estimates—in other words, best cases.

Planting trees

(-) 5 tons CO_2e 100 square meters (330 square feet) of replanted broadleaf forest in the UK

(-) 500 tons CO_2e per hectare (2.5 acres) of replanted broadleaf forest in the UK

(-) 200 billion tons CO_2e total carbon storage of planting 900 million new hectares (2.25 billion acres)[1]

> Tree planting works (and we can do it as individuals) but it's a limited solution

Tree planting is the simplest of the more natural solutions. If we do it well, we can also use it to deal with the biodiversity crisis that is going on at the same time as our climate emergency. And we can make the world more beautiful while we are at it. If we do it badly, we'll end up with hideous monocultures.

One study claims there is global potential to plant about a trillion trees around the world, which could absorb around 200 billion tons of carbon dioxide over a hundred-year period. That's equivalent to around six years of today's CO_2 emissions (although some scientists have argued that these estimates are too high).[2] You also need to remember that planting trees is largely a one-off win. Once the forests have matured, their rate of carbon capture becomes much less. So, if we rely on this card without cutting our emissions, we will be in trouble.

If you decide to use tree planting as your own personal offset, you should reckon on planting 1,100 square feet (that's half a tennis court) of forest each year to balance a 5-ton footprint, based on the carbon sequestration from UK forests. That's for each of us. As I write this, high-quality forestry offsets in the UK can still be bought for around £20 ($25–30) per ton. That's an unrealistically cheap carbon price, reflecting the sad fact that there are still low-hanging fruit that urgently need picking. In recognition of this we might think about doubling our offsets or more. And putting some money, on the side, into protecting tropical rainforest. If you decide to do any of this, Woodland Trust (and their international partners World Land Trust) and World Wide Fund for Nature are reliable options, and Cool Earth is a good charity for rainforest protection.

Marine planting

(-) 420 kg ((-) 924 lbs) CO_2e per hectare (2.5 acres) per year for replanted seagrass after 15 years[3]
(-) 37 billion tons CO_2e restoring seagrasses, mangroves, and salt marshes to pre-World War II levels[4]

> The underwater version of tree planting

In 2020, Sky Media announced it would be planting seagrass in areas around the British coast where the coverage has plummeted in recent years as part of its target to go net zero. Seagrass only covers 0.2 percent of the seafloor but absorbs 10 percent of the ocean's carbon each year. This may indeed be a very worthwhile thing to do and it has the backing of World Wide Fund for Nature. But, like forestation schemes, the scope for it is fundamentally limited and we need to do it anyway.

Soil carbon sequestration

(-) 65 billion tons CO_2e global maximum feasible soil carbon restoration over a 20-year period[5]

> Better farming practices can make a big difference

The idea here is that you change an agricultural practice in such a way that the soil captures more carbon. If you can get the complex science right, you might get a few tons of carbon saving per hectare for a few years and then you more or less reach a limit. And then you have to maintain the practice; otherwise, the carbon comes back out again. So, you are committed to carrying on forever without much in the way of continuing carbon savings. Even the number quoted here has been called into question, with some claiming that using soil carbon sequestration as a mitigation tool is unfeasible.[6]

Despite its scientific uncertainties, fundamental limits, and permanence questions, soil carbon sequestration practices look very worthwhile, as long as the ones we adopt are consistent with feeding the world and enhancing our biodiversity. I am particularly wary of some of the claims made about the potential to do this through the right kind of cattle farming (as popularized by Allan Savory), as I've seen a few wild over-claims but nothing that really results in carbon-friendly food production.

Biochar

(-) 15 tons CO_2e per hectare (2.5 acres) of biochar[7]

(-) 1.8 billion tons CO_2e annual global carbon sequestration potential of biochar[8]

> Carbon can be more or less permanently sequestered as charcoal

Biochar, the spreading of charcoal on fields to capture carbon, is really just a variation of soil carbon sequestration. The storage here isn't quite permanent, but it might last for a few hundred years, which is probably good enough. It is another option with finite capability, but it has the added advantage of enhancing soil fertility and thus helping with food and land challenges. A few genuine offsetting operations, such as Puro.earth, are offering individuals and companies the option to invest in biochar projects, and they are worth supporting as an emerging tool to combat climate change.

BECCS (bioenergy with carbon capture and storage)

(-) 12 billion tons CO_2e annual potential storage[9]

> Burning biomass and storing the emissions

Carbon capture and storage (CCS) at the point of combustion doesn't really count as removal, but the technology is just about ready to go, whenever the funding arrives. It will be useful while fossil fuel persists in the energy mix, but since it only deals with big sources of combustion—steel plants, for example—it will only ever be capable of capturing a modest proportion of our emissions.

For carbon removal, a step on from CCS is the idea of BECCS, which is to grow a biofuel and burn it, capture the emissions, and store them forever. This concept was developed as an emergency measure to have up our sleeve in the event of catastrophic climate impacts, but it has somehow found its way into mainstream climate mitigation pathways.

A disadvantage of BECCS is that it puts pressure on our food and land system because the land used for its biomass can't be used for growing food or be allowed to grow wild for biodiversity. On top of that we haven't demonstrated it at scale yet, and it comes with the risk that the methods for locking the carbon away turn out not to be as permanent as we'd hoped, which at worst could be disastrous. And, like all natural carbon sink methods, it has a finite limit, determined by both land requirements and storage capacity.

Enhanced rock weathering

(-) **1.1 billion tons CO_2e** per year current global carbon sequestration by natural weathering

(-) **5 billion tons CO_2e** per year potential carbon capture using basalt[10]

> Mimicking a natural process

Over a billion tons of CO_2 is already sequestered each year by nature through rock weathering. In an enhanced process, finely crushed basalt or dunite rock is spread on fields as a powder, to a depth of around 8 mm, where it absorbs CO_2 at a faster rate.

One recent study estimates the cost of this at just $60 per ton if dunite is used, with a possible potential for absorbing a massive 95 billion tons of CO_2 per year.[11] That is more than twice current global emissions. But unfortunately, there are problems. One potential show-stopper is that dunite has traces of harmful minerals such as chromium and nickel, which could leach into soils in a bio-available way and make agricultural land unusable. Then there is the risk to human lungs from creating so much fine powder and spreading it all over the world. Finally, there is the scale of the mining operation to consider. To take just 1 billion tons of carbon out of the air per year would require mining operations on almost the scale of the world's current iron ore extraction.

Alternatively, at $200 per ton of CO_2, basalt can be used, and this has the added benefit of improving soil fertility by adding potassium. However, studies estimate that the capture potential using basalt is much reduced, at just under 5 billion tons CO_2 per year—well worth having, but not a game-changer. Every ton of carbon removed from the air requires around 4 tons of basalt.

Rock weathering, however, avoids the risks associated with deep storage of CO_2 underground or at sea, as well as the potential land use loss of mass sequestration through monoculture forestation.

DACCS (direct air capture and carbon storage)

(-) 12 billion tons CO_2e per year[12]

> Artificial trees—on a grand scale

Direct air capture of CO_2 is a long-proven technology. Submarines have used it for decades to prevent a dangerous build-up of the gas. The holy grail is doing it at scale and at low cost in terms both of energy (it makes no sense to generate emissions in removing CO_2) and expense. And, of course, there is the problem, once captured, of long-term storage of the CO_2. But, in theory at least, DAC could be hugely important.

There are still technological uncertainties. We don't know for sure whether an injection of serious funding would lead to us being able to scale up as fast as we need and, as with BECCS, DACCS carries challenges in finding means of carbon storage (old oil wells are the most obvious) and that they may turn out not to be permanent.

In 2013, I wrote in my book *The Burning Question* that it would be unwise to rely on such an uncertain emerging technology. Today, while I remain uncomfortable, it is clear that we need to pull every lever we can, including DACCS.

There are two main players, currently, in direct air capture: Climeworks (based in Geneva) and Carbon Engineering (based in British Columbia). Carbon Engineering is backed in part by Bill Gates, and Microsoft has pledged not only to go carbon zero by 2030 but to "remove enough carbon by 2050 to account for all its emissions since its founding in 1975." The two companies operate somewhat different technology. Climeworks has perhaps the most imagination-capturing prototype, which pulls CO_2 from the air in Iceland, mixes it with water, and pumps this into formations where, in a matter of months, it reacts to become rock (calcium carbonate). It is possible for individuals or companies to invest in this genuine offsetting, though at a sizeable cost of nearly $1,000 a ton.

Both Carbon Engineering and Climeworks suggest that the price for CO_2 removal and storage could drop to below $200 a ton within two or three years. That would make DACCS a significant technology. But while in theory there may be a limit on the amount of carbon that DACCS could sequester, in practice there may be huge problems with scaling up both the capture and the safe storage. We should not think of this as a silver bullet.

What can we do?

The climate emergency is a global challenge that demands a global solution and it is easy to feel that our own individual action is of no consequence. It's easy, too, to fall for the line that is often spun by politicians opposed to taking climate action, that sacrifices made by the US make no difference, as it won't make any difference to CO_2 emissions in China, for example, or other parts of the world. But that's not a reasonable response. The US is responsible for 12 percent of global emissions; cutting those in half and driving to net zero has a massive impact, both of itself and as a global pathway. And each one of us, as individuals, can be a meaningful force for change, both by cutting our personal carbon and exerting pressure on our workplaces and schools, our councils and governments, and on companies and corporations.

When I wrote the first edition of *Bananas*, I was keen not to tell anyone what to do. I don't particularly enjoy receiving instructions on how to live, and I figure that most people are the same. But the first question I was asked at every book talk was invariably the title of this chapter: *What can we do?* So, this section is my attempt at an answer, or at least some suggestions, based on fifteen years of thinking, talking, researching, and consulting on carbon footprints, and plenty of practice in not being quite as good as I'd like to be at walking the talk.

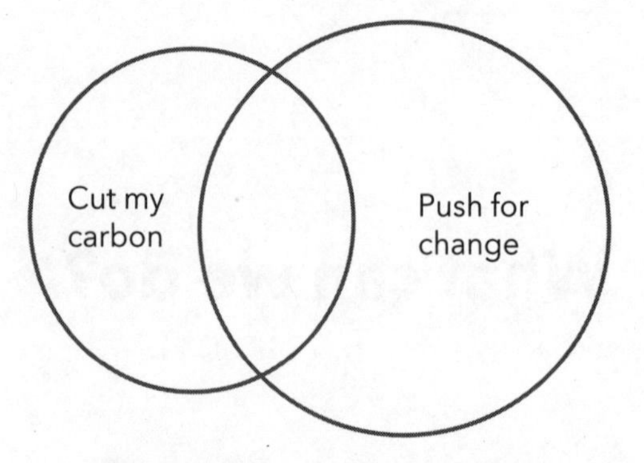

Two essential and linked areas of action,
drawn to scale in terms of their importance

This section is in three main parts. In the first, I will talk through why individual actions are still so important and effective despite the global nature of the challenge. In the second part, I'll look at the details of practical carbon cutting. Then, perhaps most important of all, I'll address the question of how we can exert our influence and push for the big changes that we need. Cutting our individual carbon footprint is essential, but the other ways we can push for change are even more powerful, as I hope the above diagram shows.

Even though none of us are going to be able to show how we single-handedly transformed the whole world, we can become part of a movement, the power of which is far more than the sum of the parts. And that's enough.

Why individual action matters

Here's the first reason. Even though each of us is such a small part of the change that is needed, if we take action and consciously live a lower-carbon lifestyle, we help to create new norms. By finding ways to live better and with less impact, we show others what is possible. A humbling aspect of human nature is that most of us like to be like most other people most of the time. So, anything you do that cuts your carbon makes it more normal for others to do the same. If you do something that feels awkward or unusual, take heart from the knowledge that you are making it easier for other people to do it. If you pick up on a sustainable trend, you create permission (and then, in time, pressure) for others to do likewise. Wasting our remaining carbon budget will become seen as irresponsible and foolish.

In the same way, through making changes in our own lives, we open up political space and create a market for more sustainable business ideas. It seems clear to me, for example, that in 2019 striking schoolkids and Extinction Rebellion (XR) activists made it far easier for the UK to tighten its carbon targets (in June 2019, the government amended the 2008 Climate Change Act to commit to zero carbon by 2050, closer to where it needs to be) and changed the conversations in the businesses I work with. Individuals changing their diets are undoubtedly encouraging vegan options in all food chains. By making our own changes, we demonstrate that we care. It is our way of proving to the politicians and businesses that we are serious. In the diagram opposite, this is where the two circles overlap.

Personal carbon cutting also brings integrity to all the other ways in which we need to push for the low-carbon world. And by going on the low-carbon journey personally, we learn firsthand about the global issues. The things that hold us back personally (cutting our flights, for instance, or changing our diet) have parallels at national and international levels. The negotiations with family and friends teach us the difficulty of international climate talks. We shouldn't expect it to be easy to make sustainable choices, and we don't have to beat ourselves up for not being perfect. We should be curious about our failings.

How can I cut my footprint?

Everyone's footprint has a different size and shape. In order to pick your battles effectively, it is a good idea to know a little bit about what your own footprint is like. You don't need a perfect analysis; just a rough idea of what the big deals are.

STEP ONE: Understand your own carbon

In the US, the average person's footprint is about 21 tons CO_2e and looks like the pie chart opposite.[1]

This includes a domestic flight about once a year and owning a gas- or diesel-powered car (between two people) which does about twenty-five miles to the gallon, on average, and is driven about 13,500 miles per year. The average home goes through just under 4,000 kWh of electricity and 4,000 kWh of other fuel per person per year. The average diet includes eating meat most days and throwing out about 32 percent of the food (by weight) that is bought. Quite a bit more clothing, furniture, and other stuff is bought than is really needed and quite a bit of it ends up thrown out well before its time.

Of course, there is no such thing as an average person. The US is an even more unequal society than the UK, with 40 percent of the wealth controlled by just 1 percent of the population—whose combined wealth is, in turn, greater than the combined wealth of the poorest 90 percent of Americans.[2] Correspondingly, a small number of people have a much higher footprint than average. Partly for this reason, over half the US population never takes a flight. If you're among the frequent flyers, then you have some big potential wins in getting your footprint down to the average, and toward the 5-ton lifestyle that seems a possible target for us all in the next few years.

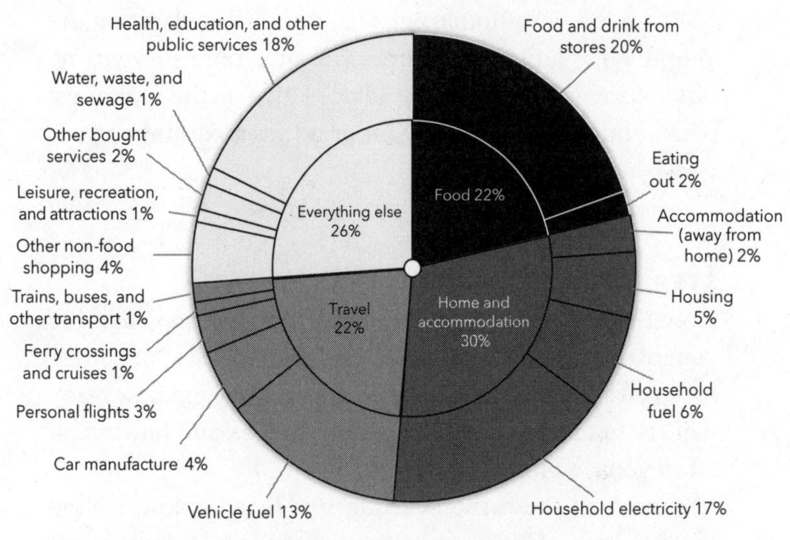

The average US person's annual footprint of 21 tons CO_2e per year[3]

Based on the average footprint and your knowledge of how you compare to that, you can probably create a rough pie chart of your own footprint. I have even left a space for you to do this here if you want to. For example, if you fly more than the average US citizen, your personal carbon footprint should have a larger share coming from flights than in the figure above. If you eat meat only once a week, your food and drink will have a smaller share. (Tip: Use a pencil.)

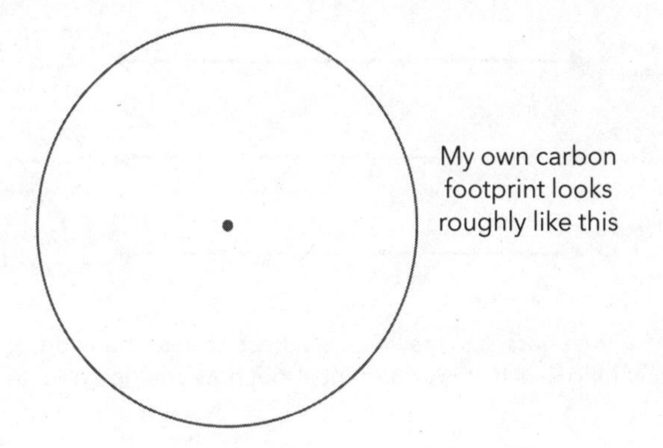

There is a very simple personal carbon calculator on the World Wide Fund for Nature's website: footprint.wwf.org.uk/#/. I am going to base the advice I give on the typical UK person, but you can adjust your priorities according to your own personal pie chart.

STEP TWO: Pick your battles

Now that you have a rough idea of the size and shape of your own footprint, you are in a position to pick your battles, based on a blend of the size of each part of the pie, how ready you are to make different types of changes, and how much effort you are prepared to make.

Across all the carbon-cutting tips listed below, I hope there is something for everyone. Low-carbon living isn't only for the poor or the rich, or the busy or those with time on their hands. It is for everyone, but there are many different ways of going about it. So, it's worth thinking about the *style* you want to adopt. One way to do this might be to position yourself on the scales below. There might be other considerations as well, such as any impacts on your social life that you might seek or want to avoid.

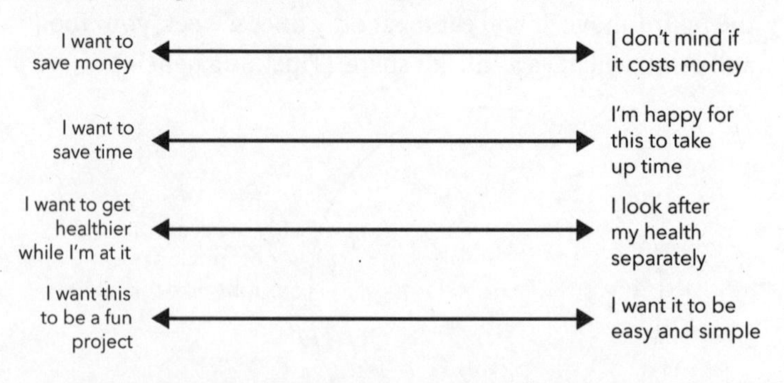

Put yourself somewhere on these scales. Then add anything else that is important about how you go about this.

STEP THREE: Get started on pieces of the pie

What follows is a tour of the pie and a raft of ideas. As you consider these, remember that this is not about making yourself miserable or beating yourself up for your shortcomings. We are where we are, and the key is to move forward. As with any diet, some habit changing will be needed, but overall think of this as an opportunity for making life *better* than it was before. Sometimes, a bit of creativity will lead to a win on every front. Other times, the only reason for an action is because we know it's the right thing. However, if an action leaves you with a bad feeling, don't do it. But don't give up the challenge, either. Keep it in your mind over time, looking for a way forward that genuinely works for you.

As you go through, put a check mark, X, or question mark next to each of the ideas if you like to indicate which ones you might be up for.

Food

Food is around 22 percent of the average US person's footprint. The following simple rules, starting with the most important, will cut half to three-quarters off most people's food footprint:

- **Eat less meat and dairy** and especially less beef and lamb. If you do buy these meats, make sure they are from mainly grass-fed animals that are not on deforested land and not on land that should be used for crops (good options might include hill sheep and cows). Try the ever-more-tasty veggie alternatives and, as lab-grown meat becomes available, choose this in preference to animal-factory meat.
- **Eat everything you buy** Check what needs eating before going to the store, learn to love leftovers, and keep vegetables in the fridge (except potatoes). Freeze things before

they go off. Give away food before it goes to waste. Buy only what you know you can eat. Be especially cautious of buy-one, get-one-free offers on fresh foods.

- **Avoid airfreighted food** If fruit or vegetables have come a long way, ask yourself whether they are robust and thick-skinned enough to go on a boat, or whether they will have had to be flown. It should say so on the label, but as it often doesn't, here are some examples: Apples, oranges, bananas, pineapples, and melons should be fine any time. Any vegetables from Central or South America are usually not. Buying a lot of airfreighted food can raise a vegan diet to the footprint of a meat-eater (see p. 122).
- **Try to reduce packaging** Take reusable bags to the grocery store or retail stores. Buy reusable mesh bags for produce. As of February 2021, eight states (California, Connecticut, Delaware, Hawaii, Maine, New York, Oregon, and Vermont) have banned single-use plastic bags. Some stores (Whole Foods) have banned single-use plastic bags already, and others (Kroger, America's largest grocery chain) plan to phase them out over the next few years.

This is good, but the most important considerations when shopping for food are avoiding airfreight, buying in season, and making your diet as plant-based as possible. You can find seasonal fruit and vegetables in your state by using the following website: seasonalfoodguide.org.

Travel

Travel makes up another 22 percent of the average person's footprint in the US. The biggest deal is driving, followed by flying. And a luxury cruise is the biggest carbon expense possible.

Driving

- **Ditch the car** This is the ideal option. Kiss goodbye to parking strife on your road. Save a ton of money and live a healthier life. Rent a car when you absolutely need it. Or join a ride-sharing group. You probably won't miss it.
- **Drive less** Take public transit whenever you can. Cycle and walk. Get an electric bike (see p. 30) because these use a tiny amount of electricity compared to a car (typically about one-twentieth) and make it easy for you to travel twice the distance that you might be prepared to do on a conventional bike. And you still get the health benefits. In fact, there is some evidence that people with electric bikes get more exercise than people on normal bikes, because they do so many more miles on them. A nice electric bike can be a bit pricey, but so much better than being stuck in traffic all the time.
- **Share the journey** This is also more sociable and cheaper.
- **Drive carefully** Slowing down from 80 mph to 55 mph can take about a third off your fuel consumption. If you are heavy on the pedals in urban or rural driving, you can cut about a third off your fuel there too, just by reading the road better and adopting a more sympathetic approach. Do not accelerate and brake between speed bumps. All these steps also make you safer, save money, cut your stress levels, and keep the air cleaner for everyone.
- **Think before buying a new car** To reduce the embodied carbon in the manufacture of the car, keep your car on the road for as long as you can. The exceptions to this are if the car is hopelessly inefficient or you have to do very high mileage (over 20,000 miles per year, say), in which case a modern, more efficient replacement will be better. If you do need to buy a new car, go for a small, efficient one, ideally second-hand, and, if you can, make it an electric one or a plug-in

hybrid. And don't think that because it's electric you don't need to worry about the carbon footprint, because the electricity has to come from somewhere and it will have taken a lot of energy and carbon to manufacture the car in the first place.

Flying

At 3 percent of an average US carbon footprint (and rising fast), flying is for many of us an uncomfortable area in addressing our carbon footprint. The hard truth is that we are a long way off working out how to put a long-haul passenger aircraft into the sky without burning through something in the order of 100 tons of fossil fuel. This produces about three times its weight of CO_2, and the effect of those emissions is almost doubled by high-altitude factors, so there is no getting around the need to be frugal with our flying. Don't believe anyone who tells you otherwise.

Here are some tips to help you cut down on flying:

- **Take the train** Trains take longer and are often more expensive than flights (booking well in advance helps), but they are also super-relaxing and typically only one-tenth of the carbon footprint. You can get from London to Barcelona by train in a (fairly long) day. But try thinking of a European train journey as part of your holiday, and have fun figuring out where you might stop en route.
- **Have a video conference or phone call** instead of attending a meeting or a conference in person. This saves plenty of time and money, as well as a lot of carbon (see p. 27).
- **Holiday closer to home** This will give you an extra day's holiday at each end and you can spend the flight money on treats you wouldn't otherwise allow yourself.
- **Fly economy** If you decide your flight really is important, make sure you go economy class, because business class has

about twice the carbon footprint. You could also think about offsetting your flight in a genuine way (see p. 200). Expect to pay at least $135 per ton if you want to remove or properly balance the carbon. But do not consider this as "making it okay"; rather, it's a note to self that you understand the impact.

- **Skip the luxury cruise** If you've been contemplating going on a cruise, please think again. Travel by ferry is better than flying, but luxury cruises are not, mainly because the space requirement per person is so huge, and therefore so is the fuel use per person. Sea freight is much more efficient than flying, but for humans this would only apply if we were prepared to snuggle up like sardines rather than requiring swimming pools, ballrooms, cinemas, and, of course, our own cabins.

Home energy

Emissions from our homes represent about 30 percent of the average footprint and there is a range of actions you can take to reduce it. Some will save money as well as carbon right away and require no planning, while others may cost thousands of dollars and require a detailed technical assessment but will pay back over the years, again both in carbon and cash. Then there are some actions that will never pay you back financially, but if you can afford them, they might still be worth it for the sake of knowing you have cut your footprint. I'll start with the simplest and cheapest and work upward.

Green energy

Switch your energy supplier to one offering a green renewable tariff. Green tariffs allow energy customers to choose clean power sources such as wind or solar rather than "brown electricity," such as that generated by coal. A local utility, approved by the state's public utility commission, allows eligible customers to source some or all of their electricity from renewable resources.

But how do you choose a genuine one? Some suppliers buy up Renewable Energy Certificates (RECs), which show the percentage of the energy they provide that comes from renewable sources. But these do little to encourage a genuine shift to renewable energy. What you really need is a supplier that sources 100 percent of its electricity directly from renewable energy generators. You can find companies that do this in your state by using the "Find green e-certified tool" recommended by the EPA, which can be found at the following link: green-e.org/certified-resources. Make sure you search for "residential renewable energy," as only this option gives you companies that buy directly from renewable sources.

INSTANT MONEY AND CARBON SAVERS

- Turn off lights and appliances when not in use. This can include turning off Wi-Fi routers overnight or when on holiday.
- Turn off the heating in rooms you are not using and have it on a timer so that it is on only when needed.
- Run washing machines and dishwashers on the lowest temperature setting that does the job.
- Hang things out to dry instead of using a tumble dryer (which wears your clothes out faster, too).
- Wear a sweater and thick socks or warm slippers if it is cold. Have blankets readily available in your living room.
- Use a hot water bottle.

- Boil only the water you need in your kettle (thus also saving time) and, when boiling food, do so on a gentle boil (which is exactly the same temperature and just as quick as a ferocious hard boil, but uses far less energy).
- Keep your showers short and your baths less deep, or perhaps share bathwater with loved ones. Think of a bath as a treat.

SMALL COSTS WITH LONG-TERM SAVINGS

Now for the jobs that cost money. Going down the list, the need for skilled expert advice increases.

- Buy hot water bottles.
- Fit LED light bulbs everywhere.
- Insulate your attic to a very high standard.
- Switch to a renewable electricity supplier, but don't treat this as an excuse to use more energy.
- Fit smart thermostats to radiators, so every room can be at the temperature you want and no higher.
- Maintain your furnace and replace it if it is not as efficient as it could be (unlike with cars, the embodied energy and carbon in the furnace itself is nothing compared to the fuel it uses).
- If you burn wood, make sure the stove is efficient, the wood very dry, and you are using it in the most efficient way (ensuring total combustion of the wood, but not burning so hard that all the heat is going up the flue). If you are in an urban area, think about not using it at all (because, even with the cleanest burn you can get, the particulate pollution is bad for everyone's health in the neighborhood).

BIG CHANGES

For all these interventions, you should get proper advice before you begin. Even if you end up spending $1,000 on expert advice to understand your options, it could be money well spent. The US has a particularly poor housing stock in terms of energy consumption and needs huge investment in the kinds of actions listed below.

Here are some home changes you could consider:

- Double or triple glaze your windows.
- Fit solar panels.
- Fit ground- or air-sourced heat pumps.
- Insulate external walls (being careful to manage moisture levels).
- Fit a smart heating system.

In the end, the target has to be a net zero energy and carbon home. Some of us can do even better than that. Just about all our homes have to stop using fossil fuel at some point, and the sooner the better, so once your home is energy efficient, think about switching to electricity only. Even then, don't imagine that if you use renewable electricity, or generate your own, it is okay to be wasteful. Remember that the world's energy use needs to go down.

LIFESTYLE CHOICES

We need to keep asking ourselves questions about where we live.

It is more carbon efficient to put your whole home to good use, with an appropriate number of people living in it. Downsize or rent out rooms if you are starting to rattle around, or if some of the rooms are not fully used.

If you rent, try to influence your landlord to improve the energy and carbon credentials of the property.

When buying a home, especially if it is a new one, ask closely about the energy consumption. One reason developers aren't building to top energy standards is that they don't think buyers care enough to make it worth their while.

Oh yes, and if you have an AGA oven, get rid of it as soon as you can. An AGA is a British cast iron range that you might expect to see in the wealthy country kitchen of Downton Abbey. It operates something like a huge slow cooker. But whatever the salespeople tell you (no doubt, to justify a price tag over $20,000), the AGA is a carbon nightmare. It's not just that they burn fossil fuel but that they are left on 24/7. Thus, most of the heat they put out is not needed. Even the electric ones are hopeless. In fact, they are so bad that ideally you shouldn't even sell it secondhand. It is one of the few items that is better off in a recycling bin than on eBay.

Stuff and services

This catch-all makes up the final quarter. It includes the non-edible things that we buy (except cars and homes; see above), such as clothes, furniture, appliances, and IT. It also includes private and public services like schools, healthcare, and financial services.

CLOTHES

- **Buy less often**, buy higher quality, and try to understand something about who made it and the conditions they worked in.
- **Choose recycled and renewable materials** Patagonia, for example, has a target to use only such materials by 2025.
- **Wash clothes only when needed** The fluff in the washing machine is your clothes getting worn out as you clean them.
- **Repair clothes** or have them repaired.
- **Sell or pass clothes on** or donate them to thrift shops.

- **Buy secondhand or rent**, especially one-off items like formal wear or wedding dresses.
- **Trim your wardrobe** down to only the stuff you might actually wear, so that others can wear the rest.

FURNITURE

- **Buy (and sell) secondhand** Make use of sites like Gumtree, eBay, your local secondhand stores, or freecycle.org.
- **Choose sustainable materials** As a general rule, wood is better than metal, which is better than plastic. Choose sustainably sourced wood and, where possible, recycled metal and plastic.
- **Make your own**, perhaps cobbled together from other stuff.
- **When you buy new items**, try to get things you can hand down to your children and they can pass on to theirs. Try to buy from a brand that uses sustainable materials, is built to last, and will help you repair it or sell it on.
- **Repair** everything possible.
- **Recycle everything possible**, making sure it is segregated into its components.
- **If you move house**, keep the old kitchen units, if they still work. Don't rip everything out! Move away from matching sets.

APPLIANCES

- **Repair** whenever you can (YouTube has a tutorial for just about everything).
- **Buy (and sell) secondhand** (see the section on furniture, above).
- **If you have to buy new**, look for a brand that makes appliances to last and that can be easily repaired. Always buy the most energy efficient that you can.

TVS AND IT

These days the energy and carbon your device will use during its lifetime is almost certain to be less than the footprint of its manufacture. So, the critical factor is to make things last.

- **Buy quality devices and look after them** so they last as long as possible. Try to keep your phone for five years, your laptop for ten, and your TV for more. Buy brands that will help you do this.
- **Buy secondhand or refurbished devices** and pass them on when you no longer want them.
- **Don't have more devices than you need**, including smart devices such as smart speakers and smart watches.
- **Don't go for an unnecessarily huge screen** This especially applies to TVs and laptops. And there's a fashion for using two computer screens, but do you really need them?
- **Buy devices that are more ethically and sustainably produced** and can be repaired and upgraded easily, such as Fairphones.
- **Use Wi-Fi** instead of mobile data, especially for large downloads. This is a small action but worth a nod.
- **Emails** are not a problem, despite media stories. There's no need to cut them down if they are genuinely useful (see p. 17).

TOYS

- **Buy fewer toys**, concentrating on the ones that allow kids to improvise and invent.
- **Join a toy library** This will also save you money and gives your kids a better choice of high-quality toys.
- **Pass on stuff** to your friends and community and sell major items on the usual websites.

HOBBIES

- **Take up exercise and do it locally**, like walking, cycling, running, almost any other sport, or gardening.
- **Try making things** from scratch and look at joining a local community repair club.
- **Consider volunteering** to do things such as tree planting.

MONEY

Every time we spend or invest money, we support one version of the future or another.

- **Change your bank** to one that invests sustainably. A few examples include Amalgamated Bank, Beneficial State Bank, and Clean Energy Credit Union. You can find banks and credit unions that don't invest in fossil fuels at mightydeposits.com/posts/environmentally-friendly-banks.
- **Change your pension plan** to one that does not invest in fossil fuels. In 2019, my company, Small World Consulting, set out to find a plan. A scheme with Royal London Exchange came out on top. Aviva was a runner-up. You can also find out if your retirement plan invests in fossil fuels and look for some better alternatives at fossilfreefunds.org.
- **The Make My Money Matter campaign** is one way you can help to apply pressure on pension funds and banks: makemymoneymatter.co.uk.

STEP FOUR: Pick the actions you are going to start with, including at least one big one, and start pushing for change

In parallel with the carbon-cutting side of things, it's important to push for the big system changes we need. It's not one or the other; we need to do both and, in fact, they tend to feed off each other. All of us have different opportunities to influence. So, the first step is for you to reflect on your own biggest areas. These might include:

- With family and friends
- Where you work or study
- Where you shop and for what
- How you express yourself politically

WITH FAMILY AND FRIENDS

This can be the most challenging area. All of us know what it's like when a friend or family member is coming from a different position on an important issue. And it isn't possible for all of us to be in the same place on the environmental emergency and how we need to take appropriate action. We have to find clever ways of moving ourselves and each other in the right direction that bring us closer to the people we care about the most.

WHERE YOU WORK OR STUDY

Whatever our position, all of us can make suggestions, help set the culture of a business or organization, and challenge what isn't right. If one person does this, it makes it easier for everyone else to have a say. Businesses need to ask themselves what their role is in the world and whether their impact is positive. If it isn't, it needs to change. It can be hard to ask the awkward questions of the people who pay our salaries, but if we can't we have allowed ourselves to be bonded laborers. We need to bring our whole selves into the workplace, not just the commercial bit of ourselves. And we have a duty to help others do likewise. Try to look for like-minded people within your workplace or place of study and join forces with them or start a movement yourself. This can be taking part in school climate strikes, putting pressure on your university to divest itself from fossil-fuel investments and to have a bold sustainability strategy in place, or pushing for your workplace to prioritize reducing its carbon footprint. If you have a more senior role, you have a bigger responsibility for setting a culture in which the big questions can be asked and taken seriously.

WHERE YOU SHOP AND FOR WHAT

We've gotten used to a world in which it is very hard to know what goes on in the supply chains of goods that we buy.

Ask yourself what the carbon footprint might be. Ask about all the other things that might matter, from labor conditions to biodiversity. Make it a habit not to buy anything without at least asking yourself the question.

Try to build up your knowledge of more and less sustainable brands and products. One very good source of information you could try is *Ethical Consumer* (ethicalconsumer.org).

HOW YOU EXPRESS YOURSELF POLITICALLY

This goes beyond traditional party politics. We need honest, coherent politicians taking strong action. They all need to hear from us that we care. We need to get this message out, not just at the ballot box, but well before any elections, while they are still working out their messages and policies.

- **Talk or write to your Member of Congress** If an election is coming up, make sure all candidates know you care enough about climate change that it will swing your vote. Traditionally, very few people do this, so if more of us start doing so they will notice the difference.

- **Question the media** Support the most trustworthy sources and treat the others with extreme caution or ditch them altogether. (A shameless plug: there is a whole section called "Values, truth and trust" in one of my other books, *There Is No Planet B*, which in many ways is the companion to this volume.)
- **Vote** for politicians who genuinely "get" the climate emergency and talk in a coherent way about dealing with it. If none of them meet the criteria, vote for whoever gets closest. Under no circumstances vote for a politician who does not believe in climate change and treat with extreme caution any politician who has let such a position go unchallenged.
- **Consider protest**

I write as someone who doesn't instinctively take to the streets. But we have to get the change. We have had decades of asking nicely but (except for a COVID-19 blip) the rate of emissions is still going up rather than down. Now we have a proper emergency on our hands. So, if protest is what it takes, then it is worth it. Extinction Rebellion (XR) and sister movements, including school strikes (Fridays for Future), have had an amazingly positive impact around the world. They've been clever most of the time. They have also made one or two mistakes. I respect them hugely. For what it is worth, here is what I think is important for XR and all other protests going forward:

- **Beyond non-violence, we need to be respectful** of absolutely everyone, right down to dishonest politicians and oil executives. Absolutely everyone.

- **We need to push for truth as a value**, cultural norm, and political "must have," not just about climate, but about everything.
- **We need to give everyone a taste of a better world** Given the critical nature of the situation, I think there is now a place for respectful disruption in order to make a point but, overall, our message needs to be positive. To give an example: when, in April 2019, XR occupied four sites in London, activists put trees on Waterloo Bridge, created a skatepark for kids and a library, and gave out free food. There were talks and music. The atmosphere was wonderful. The air was cleaner, and London felt like a better place. There were teams of people picking up not just their own litter but other people's as well. There were repeated messages to all protesters to respect everyone and not to take alcohol (or drugs) onto the sites. The majority of Londoners and visitors experienced a better city as a result of their action. It was really quite moving. The police couldn't believe what they were dealing with, having never experienced such a warm and respectful protest. Yet it was clear and insistent.

We need to get better and better at engaging with every part of society. XR had a reasonable social mix, but it is so important to show that the transition we are pushing for is for everyone.

STEP FIVE: List actions you can take to push for the systemic changes that the world needs

Think about the different groups you can influence and then label the bubbles below with the biggest bubbles for the areas where you think you can have most influence. Now fill them with things you are going to do to push for change.

And that's it! You now, hopefully, have a plan that fits both your circumstances and your style. It involves cutting your carbon, but it is much more powerful than that. You are also exerting influence across all the most important areas of your life. It might not be perfect, but it is good enough. You have my very best wishes. And, more to the point, you have the best wishes of the rising billions of others around the world who have also woken up and are on your side.

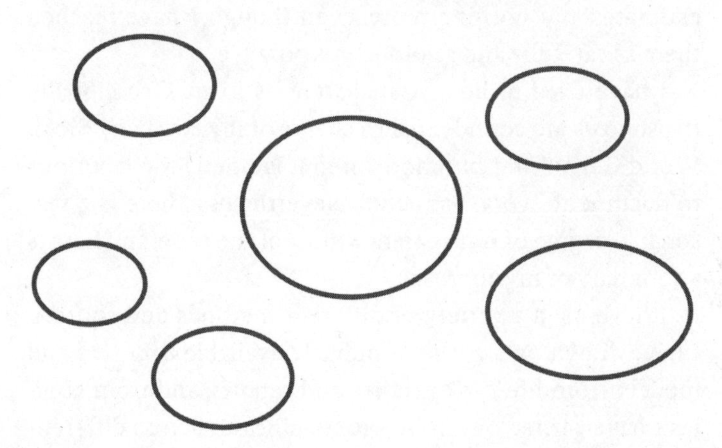

Where the numbers come from

I hope I have made the point, right through this book, that carbon footprinting is a long way from being an exact process, whatever anyone ever tells you or whatever numbers you might see written on products. All my numbers are best estimates and nothing more, even though I have reached them as carefully and rigorously as possible.

I have tried to be as transparent as I can. Occasionally the sources are confidential to clients of my company, Small World Consulting, but more often it is simply too laborious to document every last detail. Nevertheless, there is a reasonable degree of transparency most of the time, and here is a summary of my approach.

I have used a variety of different methods and sources. I have drawn on a range of publicly available data sets and models, from life-cycle studies and reports, and from studies I have carried out myself for businesses across different industries. I have used models that we are developing all the time at Small World.

Often, I've arrived at numbers from a couple of different routes to check that the results agree with each other. I've tried to put notes and references in the text wherever possible. Occasionally, frankly, it has been more a case of putting my finger in the air and guessing, but when that has been the case, I've tried to make it clear.

Following are some of the main sources I have used.

Publicly available data sets drawn from process life-cycle analyses

Process-based life-cycle analysis (PBLCA) is the most common approach to carbon footprinting. It is often referred to as a "bottom-up" approach because you start off down on your hands and knees, identifying one by one all the processes that have had to happen in order for, say, a product to be created. Then you add up the emissions from each process and that's the footprint of the product. Simple!

Except that it isn't. Not at all. It's backbreaking work and since the number of processes you really ought to account for is more or less infinite (and I do really mean infinite, not just very large), the job is never quite complete. So, you end up with an underestimate.

These underestimates are called "truncation errors" because you have to cut some of the pathways short. And the leaks are often shocking, sometimes amounting to 50 percent or more. To make matters worse, these problems are popularly overlooked, even in the development of government-backed guidelines, such as the PAS 2050 standard, which was published despite a government-commissioned study that concluded that the draft methodology wasn't fit for some of its key intended purposes.

For all the problems, and despite being hard work, process-based life-cycle analysis is still an essential source of detailed information that can't be gathered any other way. Here are some of the key sources of this type that I've used, each of which is referenced in the main text:

The UK government's Department for Environment, Food and Rural Affairs (Defra) publishes emissions factors for a range of fuels, electricity sources, transport modes, utilities, and waste. These are mostly UK-specific and don't take account of full supply chains. I use them where I can, but supplement with additions for the missing supply chains.

The University of Bath produces the Inventory of Carbon and Energy, a publicly available data set of carbon emissions factors for hundreds of materials, mainly relating to the construction industry, up to the factory gate.

The Association of Plastics Manufacturers in Europe (APME) publishes data sets of emissions factors for a wide range of plastics, based, not surprisingly, on European manufacture.

I have drawn on a further wide range of life-cycle analysis studies from all kinds of sources that we compile into data sets at Small World. This can be tricky, because they all draw their boundaries in slightly different ways and use slightly different assumptions. At its best, this has involved picking through high-quality academic studies. Where necessary, I have resorted to just about anything I can think of to get the estimates that I've been after. And at its worst, it has degenerated into "Google footprinting"—scrounging around the web, digging for numbers that could be based on almost nothing. When I've sunk to these depths, I've let you know.

Environmental input-output analysis

This is a neat alternative and complement to PBLCA. It's not as popular, perhaps because it's a bit harder to get your head around, but it's at least as robust as anything else in the murky world of carbon footprinting. It is sometimes called a "top-down" approach because it starts by looking at the whole economy from a height. It uses macroeconomic modeling to understand the way in which the activities of one industry trigger activities and emissions in every other industry.

Input-output's key factor is a piece of funky math (for which Wassily Leontief got a Nobel Prize) that succeeds in capturing the endless ripple effects in a way that is 100 percent complete. It has the further advantage that, if you know how much you spend on something, you can get an instant crude estimate of its carbon footprint. It's like a magic trick. And, just like all the best magic, it is also a bit too good to be true—the downside of input-output analysis is that the results can be ridiculously generic.

Input-output analysis is nonetheless a powerful tool, both because it doesn't "leak" and because once the model has been built it is often easy to use. The basic technique is well established. The specific model I've used is one we developed at Small World with Lancaster University, and have been refining and updating for over a decade. It draws mainly on data from the UK's Office for National Statistics (ONS). The model I've used for this book is based on a 2017 picture of the UK economy; it deals with all the greenhouse gases and employs an emissions weighting factor for high-altitude emissions. A key weakness, which I refer to from time to time and sometimes adjust for, is that it treats imports as though they had the same carbon intensity as domestic production, whereas in reality they are usually more carbon intensive.

Most of the time I have used a combination of process-based and input-output approaches to get my numbers. At their best, process-based methods can be more precise, but input-output analysis is often able to get at places that process life-cycle analysis is unable to reach. Putting the two methods together is sometimes called a "hybrid" approach, and the result is a bit like looking through a microscope and a telescope at the same time. They each show you different things and, between them, if the lenses are clean, you might end up with a passable understanding of whatever it is you are looking at.

One of the tools we have developed is a method for estimating the likely truncation error on detailed and high-quality PBLCAs of different product types. It isn't perfect, but it allows us to make a reasoned estimate of how much a phone or computer manufacturer, for example, is likely to have underestimated the total if they carried out a detailed PBLCA of one of their products.

Booths Supermarkets' greenhouse gas footprint model

For the past ten years, Small World has mapped the carbon footprint of the Booths group of supermarkets and its supply chains, refining and updating continually. The model we now have draws on a great many life-cycle studies of foods up to the farm gate, often using those funded by Defra. Reports and agricultural models from Cranfield University deserve a mention because I've used them extensively, even though they are contentious. Also well worth a mention are five reports produced by the Food Climate Research Network, as well as a paper in *Science* by Joseph Poore and Thomas Nemecek.

The Booths model includes transport, processing, packaging, refrigeration, and the supermarket chain's other operations. All of these components are attributed to products, broken down into seventy-five categories. The model goes into a lot of detail, but that doesn't make it accurate. Human understanding of emissions from agriculture is still poor. The model is simply the best picture we have managed to achieve so far. Its purpose is purely practical and we think it is now good enough to work from, enabling actions to be reasonably well targeted on the hotspots. It is, I think, the most comprehensive model of the climate impacts of supermarket food in the public domain. Several papers have been published on the back of this work, including some by me with colleagues at Lancaster University, looking, for example, at the climate change impacts of dietary choice, or different options for disposing of food waste.

The footprint of IT

This significant part of the world's carbon footprint is fiendishly difficult to come to grips with. I used PBLCAs from companies coupled with manufacturers' reported data on device energy consumption. For global totals and the impact of networks and data centers per megabyte, I've used several academic papers and other reports, most notably the work of Jens Malmodin and Dag Lundén.

Direct greenhouse gas (GHG) emissions per GDP and per person for sixty countries

Note that these figures do not take account of embodied emissions of imported or exported products, or of international transport. They are simply estimates of the emissions that actually arise from each country.

Appendix: Calculating footprints

The carbon footprint of some foods

I've used Small World's food carbon models to pull together this list of selected food footprints at the checkout of a UK store. All food supply chains are different and our knowledge is far from certain, so this is just a rough guide. But it is good enough that you will be on the right lines most of the time. (If you work in catering, the numbers for wholesale deliveries would be similar.) Note that some of the numbers in this table have been adapted for the US in the main text. However, the numbers presented here should be in the right ballpark for both markets.

FRUITS	kg CO_2e per kg produce
Apples, local	0.3
Apples shipped from New Zealand	0.6
Bananas shipped from Latin America	0.7
Lemons shipped from Spain	0.9
Grapes transported by road from Spain	1.1
Melons shipped from Spain	1.2
Strawberries from Scotland	1.7
Strawberries transported by road from Spain	1.8
Berries, frozen	2.6
Dried fruits	2.9
Grapes airfreighted from South Africa	18.5

VEGETABLES	kg CO_2e per kg produce
Potatoes and root vegetables grown in the UK	0.3
Onions grown in the UK	0.5
Garlic grown in the UK	0.5
Lettuce grown in season in the UK	0.6
Broccoli grown in the UK	0.7
Cauliflower grown in the UK	0.7
Squash grown in the UK	0.7
Kale grown in the UK	0.9
Asparagus grown in season in the UK	1.1
Broccoli imported by road from Spain, Italy, or France	1.2
Spinach grown in the UK	1.2
Cucumber in season in the UK	1.3
Tomatoes (standard) in season in the UK	1.3
Sweetcorn imported by road from Spain	1.4
Avocado shipped from South Africa, Peru, or Chile	1.6
Cucumber imported by road from Spain	1.7
Avocados imported by road from Spain	1.8
Lettuce imported by road from Spain	1.8
Herbs grown in the UK	2.1
Legumes (peas, lentils, chickpeas, beans) shipped from America	2.1
Tomatoes imported by road from Spain	2.5
Peppers hothoused and by road from Netherlands	3.0
Mushrooms grown in the UK	4.1
Tomatoes (baby plum) hothoused in the UK	4.6
Asparagus airfreighted from Peru	18.5

CARBOHYDRATES	kg CO_2e per kg produce
Bread produced in the UK	1.1
Oats from Scotland	1.3
Flour produced in the UK	1.5
Pasta produced in the UK	1.5
Quinoa shipped from Peru	1.6
Pasta by road from Italy to the UK	2.0
Rice shipped from Asia	3.8
HIGH-PROTEIN VEGAN FOODS	kg CO_2e per kg produce
Tofu	1.5
Peas, lentils, chickpeas, and beans	2.1
Nuts and seeds	2.3
Quorn	4.0
NON-DAIRY MILKS	kg CO_2e per kg produce
Oat milk from the UK, unrefrigerated	0.2
Soy milk, unrefrigerated	0.4
Almond milk from California, unrefrigerated	0.6
DAIRY	kg CO_2e per kg produce
Fresh British cow's milk	1.9
British margarine	2.1
British yogurt	2.4
British cream	5.86
British eggs	5.9
British butter	9.8
Mozzarella produced in the UK	10.1
Mozzarella imported by road from Italy	10.3
Cheddar produced in the UK	11.8
Parmesan imported by road from Italy	19.1

MEAT	kg CO_2e per kg produce
Whole chicken from the UK	3.8
Whole chicken (global average)	8.1
UK bacon	10.0
UK lamb	21.0
UK beef	25.0
Imported beef steak, from deforested land	83.3

FISH	kg CO_2e per kg produce
Sardines freshly caught in the UK	2.0
Mackerel freshly caught in the UK	2.1
Prawns and shrimps freshly caught in the UK	3.8
Fresh farmed Scottish salmon	4.1
Cod freshly caught in the UK	4.1
Cod transported by sea from Iceland	4.4
Tinned tuna	5.3
Tuna airfreighted from the Seychelles	22.0
Tiger prawns farmed in Thailand	25.0

OTHER	kg CO_2e per kg produce
Cocoa powder shipped from Africa	1.9
Sugar	2.1
Local jam, honey, or marmalade	2.3
Chocolate processed in Europe and imported by road	2.4

ALCOHOLIC DRINKS	kg CO_2e per kg produce
Beer (local cask ale)	1.2
Beer bottled and brewed in the UK	1.5
Beer bottled and shipped from Italy	1.7
Wine shipped from New Zealand	2.0
Wine imported by road from France	2.3

The carbon footprint of spending money

These numbers are very generic, but do include entire supply chains, however complex. Remember that, in reality, carbon doesn't always go with cost, nor is the carbon intensity of any item always typical of the broad spend category that it is in. Also, these numbers are based on a model that assumes imports have the same carbon intensity as if they were made in the UK, whereas manufacturing elsewhere in the world can make a huge difference to the footprint per £1 (usually upward). Note that these numbers are based on GBP, not USD. For the numbers in the main text, some adaptations were made to be specific to the US.

GOODS AND SERVICES	kg CO_2e per £1 at retail prices (2018)	kg CO_2e per £1 at business-to-business prices (2018)
Textile fibers	0.18	0.33
Clothing	0.09	0.17
Leather goods	0.15	0.25
Paper	0.48	0.68
Printing and publishing	0.37	0.37
Paints, varnishes, printing ink, etc.	0.43	0.60
Soap and hygiene products	0.17	0.30
Pharmaceuticals	0.15	0.19
Rubber and plastic products	0.62	0.67
Cement, lime, plaster, and concrete	1.39	1.60
Glass, ceramic, and stone products	0.57	0.75
Iron and steel	1.88	2.32
Structural metal products	0.46	0.59
Computers and electronic goods	0.22	0.28
Electrical equipment	0.31	0.40
Motor vehicles	0.41	0.50
Furniture	0.31	0.46

Miscellaneous manufacturing	0.13	0.23
Repair and installation	0.24	0.24
Construction	0.34	0.34
Rail transport	0.59	0.59
Other land transport	0.63	0.63
Water transport	1.50	1.50
Air transport	3.76	3.76
Postal and courier services	0.41	0.41
Accommodation	0.31	0.31
Food and beverage services	0.28	0.28
Film, TV, and music production	0.14	0.14
Telecommunications	0.19	0.19
Web hosting, web portals	0.13	0.13
Financial services (except insurance)	0.18	0.18
Insurance	0.16	0.16
Real estate activities	0.05	0.05
Legal activities	0.07	0.07
Accounting, bookkeeping, and auditing	0.08	0.08
Scientific research and development	0.25	0.25
Advertising	0.11	0.11
Veterinary services	0.14	0.14
Travel agencies and tour operators	0.16	0.16
Services to buildings and landscape	0.20	0.20
Public administration	0.24	0.24
Education	0.16	0.16
Health services	0.17	0.17
Residential care and social work	0.20	0.20
Creative, arts, and entertainment	0.19	0.19
Libraries, museums, and cultural activities	0.19	0.19
Sports and recreation	0.21	0.21
Repair of computers and household goods	0.16	0.16

Notes and references

These notes and references are also posted on the book's website, howbadarebananas.com, for easy access to the web links.

Introduction

1 *An Inconvenient Truth* (2006), the documentary presented by Al Gore, (https://preview.tinyurl.com/InconvenientTruth2006) has been credited with increasing public awareness of climate change not only in the US but all around the globe. It also won Gore the Nobel Peace Prize in 2007.

2 How can that possibly be when we have had some many global agreements and national targets and almost everyone reading this book will have done at least something at some time to cut their own carbon? Surely it adds up to something! See p. 209 for a quick explanation, but for the full details you'll have to take a look at one of my other books, *There Is No Planet B: A Handbook for the Make or Break Years* (Cambridge: Cambridge University Press, 2019).

3 Intergovernmental Panel on Climate Change (IPCC, 2018), "Special report: Global warming of 1.5°C," www.ipcc.ch/sr15/.

A brief guide to carbon footprints

1 *Carbon Footprinting: An Introduction for Organisations*, published by the UK's Carbon Trust (2007), defined (on p. 1) a carbon footprint in a similar way to me, but goes on to describe "basic carbon footprints" on p. 4. These are toe-prints rather than rough estimates of footprints, https://tinyurl.com/carbon-trust2007.

2 This information comes from an Environmental Protection Agency (EPA) report from 2018; https://tinyurl.com/ghgemissions1990-2018.

3 Intergovernmental Panel on Climate Change (IPCC, 2018), "Special report: Global warming of 1.5°C," www.ipcc.ch/sr15/.

Less than 10 grams

1 Based on 82 gallons, or 310 quarts, per person per day, which is about average in the US according to the US Geological Survey report from 2015: https://pubs.er.usgs.gov/publication/cir1441.

2 40 kg CO_2e per person works out to a US annual tap water footprint of around 13.2 million tons CO_2e. This is around 0.2 percent of the total US footprint of 6.6 billion tons CO_2e.
3 Based on figures for the carbon intensity of UK water supply and treatment by the UK government's Department for Business, Energy and Industrial Strategy (BEIS). The full set of emission factors can be downloaded from the Department for Environment, Food and Rural Affairs (Defra) (2019), "Guidelines to Defra's GHG conversion factors for company reporting," https://tinyurl.com/beis-emission-factors.
4 These estimates include the emissions embodied in the device, the electricity used to run it (assuming US electricity), and electricity use in the networks, data centers, and by the Wi-Fi router. They are based on an iPhone 11 with 128 GB that is kept for two years (see *Using a smartphone*, p. 126) and a 13-inch MacBook Pro with 128 GB storage that is kept for four years (see *A computer (and using it)*, p. 140) that are connected to Wi-Fi. Doing a search over mobile networks has a lower footprint, because it does not require a Wi-Fi router; it comes to 0.11 g CO_2e per minute for the internet, compared to 0.6 g CO_2e per minute for Wi-Fi. For the spam email, I'm assuming that it's sent to so many people that the footprint of the device it's written on is negligible and that nobody reads them, so it's just the footprint of the transmission, assuming it takes 5 seconds in networks and data centers to transmit the email.
5 According to an estimate by Radicati (2018), there were 3.93 billion email users in 2019 and 294 billion emails were sent every day: Radicati (2018). "Email market, 2018–2022," https://tinyurl.com/radicati2018. Statista estimates that 55 percent of all emails in 2018 were spam, https://tinyurl.com/statista-spam.
6 Assuming all emails were read on an iPhone 11. I have used a global average electricity factor for the use phase of the iPhone and Wi-Fi router.
7 Based on a footprint of 0.05 g CO_2e per spam email (assuming a global average electricity factor for the use phase of the iPhone and Wi-Fi router) and 59 trillion spam emails per year. That's based on 55 percent of all emails being spam emails (see notes 5 and 6).
8 See note 4 above.
9 Hölzle, U. (2009), "Powering a Google search," 11 January, https://tinyurl.com/powering-google (https://googleblog.blogspot.com/2009/01/powering-google-search.html). Google used an electricity intensity of 0.67 kg CO_2e per kWh back in 2009; today, global average electricity has decarbonized a bit to 0.63 kg CO_2e per kWh, so the footprint of one search has decreased from 0.2 g to 0.19 g—not much. But, if we assume Google's data centers are twice as efficient today as they were in 2009, the footprint at Google's end is just 0.09 g.

10 According to the website Internet Live Stats, https://tinyurl.com/google-search-stats.

11 The figures are based on two iPhones being used for 30 seconds each and transmission taking place for 5 seconds. They include the footprint of manufacturing and transporting an iPhone 11 to the user and the emissions from the electricity used to power the phone (see *Using a smartphone*, p. 126), about 0.79 g CO_2e. They also include emissions from the electricity used in the mobile networks that transmit the text, at 2 watts according to Jens Malmodin, senior specialist at Ericsson and an expert in the energy and carbon footprint of networks. For 5-second transmission, assuming the text is sent within the UK using a carbon intensity of the local grid of 0.65 kg CO_2e per kWh, that's 0.002 g CO_2e. Sending a message online through an app like WhatsApp has a higher carbon footprint because it requires around 5 watts in networks that can carry mobile data and 5 watts in data centers too, according to Ericsson (2020), "A quick guide to your digital carbon footprint—deconstructing information and communication technology's carbon emissions," https://tinyurl.com/ericsson2020, and specifically their background report, https://tinyurl.com/ericsson2020-background. So that's 0.009 g CO_2e for the transmission. But, because the device plays the biggest part in the footprint of sending a message, it scarcely matters how the message is transmitted.

12 World Economic Forum (2019), "Why big data keeps getting bigger," https://tinyurl.com/weforum2019. This source estimates 18,100,000 texts being sent per minute in 2019, which is 9.520 billion per year.

13 Ofcom reports that there were 79.49 million subscribers for mobile handsets in 2018 in the UK and 73.84 billion outgoing SMS and MMS sent over mobile networks; Ofcom (2019), "Communications market report 2019," https://tinyurl.com/comms-market-report. Marketing company SimpleTexting reports that Americans sent 15 texts per day, based on Zipwhip's "State of texting report 2019," https://tinyurl.com/texting-stats.

14 PlasticsEurope's Association of Plastics Manufacturers: Eco-profiles showing emissions from production of a wide variety of plastics are available from https://tinyurl.com/plasticseurope.

15 Data on plastic bag legislation comes from the National Conference of State Legislatures at the following link: https://tinyurl.com/USplasticbaglegislation.

16 Berners-Lee, M. (2019), *There Is No Planet B*. See the chapter "How much plastic is there in the world?" (1st ed., p. 55).

17 Ten seconds of drying at 1.6 kW equals about 0.003 of a kWh. The emissions from US electricity are 0.65 kg CO_2e per kWh, so a Dyson Airblade is roughly 3 g CO_2e. Using the calculation for the 6 kW hand dryer for 15 seconds gives us 0.033 of a kWh, which equals 22 g CO_2e.

18 I am assuming that this low-grade paper comes in at just 1 kg CO_2e per kilo.

10 to 100 grams

1 The UK government's Department for Business, Energy and Industrial Strategy (BEIS) gives a figure of 1,000 kg CO_2e per ton of mixed paper in landfill. The full set of Defra (Department for Environment, Food and Rural Affairs) BEIS emission conversion factors for 2019 can be downloaded at https://tinyurl.com/beis-emission-factors-2019.

2 For the number up to the farm gate, I've used a 2008 report by Defra (2008), "Final report for Defra project FO0103: Comparative life cycle assessment of food commodities procured for UK consumption through a diversity of supply chains," https://tinyurl.com/defra-fruit. I have then added on the processing and transport number using data from our work with Booths Supermarkets: Berners-Lee, M., Moss, J., and Hoolahan, C. (2014). "The greenhouse gas footprint of Booths." Small World Consulting, https://tinyurl.com/booths-footprint. Apple weight is assumed to be 112 g.

3 Saunders, C., Barber, A., and Taylor, G. (2006) *Food Miles—Comparative Energy/Emissions Performance of New Zealand's Agriculture Industry*. Research Report no. 285, Lincoln, New Zealand: Lincoln University, https://tinyurl.com/saunders2006.

4 Blanke, M., and Burdick, B. (2005), "Food (miles) for thought-energy balance for locally-grown versus imported apple fruit." *Environmental Science and Pollution Research*, 12(3), 125–127. Referenced in Defra (2006), "Environmental impacts of food production and consumption," https://tinyurl.com/defra-food, p. 47.

5 For the energy requirement, I've used the model laid out by David J.C. MacKay in the technical chapter at the back of his 2008 book, *Sustainable Energy—Without the Hot Air*, which can be downloaded for free from www.withouthotair.com/. I've added the effect of hills. Onto this, I've added the embodied carbon in the battery based on 190 Wh per kg of battery and 1,000 charge cycles over its lifetime (data from Bosch).

6 Estimates for the embodied carbon per kWh of batteries vary and depend especially on the carbon intensity of the energy used to manufacture them. I've taken a fairly central figure of 100 kg CO_2e per kWh of battery capacity based on the following literature review:

Hausfather, Z. (2019, May 13), "Factcheck: How electric vehicles help to tackle climate change." Carbon Brief, https://tinyurl.com/hausfather2019. So that is 30 g CO_2e of embodied carbon in the battery per kWh of energy transmitted to your bike. Based on my model (see previous note), you need 15 Wh per mile if you weigh 80 kg/175 lbs (with the bike), travel at

12 mph with five stops and 20 m of climbing per mile. That comes to roughly 0.5 g CO_2e per mile for the battery.

7 Peace, R. (2019), "A guide to e-bike batteries," *We Are Cycling* UK, 15 February. Cycling UK, https://tinyurl.com/e-bikes.

8 The sums: a 5-square-meter doorway, fully open for 15 seconds, wind speed through the door of 1 m per second, temperature difference of 15°C, heat capacity of air 1.2 kilojoules per cubic meter, heat supplied by gas at 0.22 kg CO_2e per kWh.

9 BREEAM (Building Research Establishment Environmental Assessment Method) is a sustainability assessment method for infrastructure projects and buildings. It is the British equivalent of North American building assessment systems such as LEED (Leadership in Energy and Environmental Design) and the Green Globes. See www.breeam.com.

I understand that the BRE has since improved its energy efficiency criteria somewhat. The sums here are based on a temperature difference of 15°C (typical for winter) and a wind speed of just 2.5 mph flushing warm air out of the building.

10 Quick, D. (2008), "Revolving door generates its own power," 12 December, New Atlas, https://tinyurl.com/rotating-doors.

11 For my calculations, I have used the exhaust emissions from the UK government's BEIS and added on the supply chain emissions and embodied emissions for the gas- and diesel-powered buses. For the electric bus, I have used BYD-ADL Enviro200EV as an example, which has a capacity of 90 people and uses roughly 1.34 kWh per mile. Multiplying the US electricity figure of 0.65 kg CO_2e by 1.34 kWh per mile gives a number of around 10 g per passenger mile. I've worked out the embodied emissions to be a further 1 g per passenger mile to get the final number of 11 g. The full set of emission factors can be downloaded from the BEIS website, https://tinyurl.com/beis-emission-factors.

12 Data comes from a recent paper: Dedoussi, I., Eastham, S., Monier, E., and Barrett, S. (2019), "Premature mortality related to United States cross-state air pollution." *Nature* (578), 261–265, https://tinyurl.com/USair pollutiondeaths, and Royal College of Physicians (2016), "Every breath we take: The lifelong impact of air pollution," https://tinyurl.com/pollutiondiesel.

13 In our input-output model of the greenhouse gas footprint of UK industries, sports goods typically have a carbon intensity of around 210 g per £1 ($2–3) worth of goods at retail prices. If we make the very broad assumption that cycling goods are typical of this, and if we say that Her Majesty's Revenue and Customs (HMRC) is being roughly fair to reimburse you at 20p (25–30 cents) per mile for business travel on a bike, then we would need to add about 42 g CO_2e per mile to take account of the wear and tear on your bike, your waterproof gear, lights, helmet, and so

on. However, there are so many variables that I have gone for a range of 10–100 g. Actually, as someone who is frequently cycling between offices and train stations trying to keep jacket, tie, and laptop dry, I suspect that HMRC has underestimated it and should be paying out the full 40p (50–60 cents) per mile that they allow for car users, which would also provide a beneficial incentive.

100 to 500 grams (3.5 to 17.5 oz)

1 For the footprint for bananas and oranges, I've updated the numbers from a piece of work I was involved in a few years ago: Hoolahan, C., Berners-Lee, M., McKinstry-West, J., and Hewitt, C.N. (2013), "Mitigating the greenhouse gas emissions embodied in food through realistic consumer choices." *Energy Policy*, 63, 1065–1074, https://tinyurl.com/hoolahan2013.
2 There is more on this on the website of the non-profit Banana Link: www.bananalink.org.uk. For a critical and pessimistic look at the future of bananas in our lives, see also Koeppel, D. (2008), "Yes, we will have no bananas." *New York Times*, 18 June, https://tinyurl.com/koeppel2008.
3 The charity Waste & Resources Action Programme (WRAP) estimates that people in the UK waste 22 percent of the food they purchase: WRAP (2020), "Food surplus and waste in the UK—key facts," https://tinyurl.com/wrap2020. A recent paper estimated that for the US this is 31.9 percent: Yang, Yu., and Jaenicke, E., "Estimating food waste as household production inefficiency." *American Journal of Agricultural Economics*, 102(2), 525–547, https://preview.tinyurl.com/USfoodwaste.
4 The number for disposable diapers comes from: Cordella, M., Bauer, I., Lehmann, A., Schulz, M., and Wolf, O. (2015), "Evolution of disposable baby diapers in Europe: Life cycle assessment of environmental impacts and identification of key areas of improvement." *Journal of Cleaner Production*, 95, 322–331, https://tinyurl.com/cordella2015.
5 The study I've used for reusable diapers is Aumônier, S., Collins, M., and Garrett, P. (2008), "An updated lifecycle assessment study for disposable and reusable nappies," Science Report no. SC010018/SR2, UK Environment Agency, https://tinyurl.com/aumonier2008. Since the first edition, I've calculated that washing at 60°C and drying on the line is now 35 percent less carbon intensive than it was ten years ago and washing at 90°C and tumble drying is 42 percent less carbon intensive, due to the UK electricity mix being a lot better than it used to be. I have factored this in to get the new numbers.
6 Hoolahan et al. (2013). See note 1, above.
7 Emissions per mile come from the US Environmental Protection Agency's *Emission Factors for Greenhouse Gas Inventories* 2019; https://preview.tinyurl.com/EPAemissionfactors.

8 David J.C. MacKay lays out the math nicely in *Sustainable Energy—Without the Hot Air* (2009), published by UIT Cambridge Ltd and available as a free download from www.withouthotair.com.

9 Kemp, R. (2007), "Traction energy metrics." Rail Safety & Standards Board, London.

10 Figures include use-phase electricity assuming the device is used in the US, a share of the carbon embodied in the production and transport of the device (assuming a lifetime of four years), a share of the emissions from standby and the electricity used in the transmission of one hour of content, and a share of the embodied footprint of the set-top box (for watching on a TV) or Wi-Fi router (for watching on a laptop). The transmission includes use-phase emissions from playout and coding and multiplexing, networks, satellites, data center storage, access network equipment in home (such as set-top boxes and Wi-Fi routers), and personal video recorders.

I allocated standby and embodied emissions based on five hours' viewing per day. In the UK, Ofcom estimates that "In 2018, individuals watched a total of nearly five hours of audio-visual content, per person per day, across all devices." This includes twenty-six minutes for subscription video-on-demand (such as Netflix or Amazon Prime), Ofcom (2019), "Media nations 2019: Interactive report," https://tinyurl.com/ofcom-media-nations.

Assuming that 80 percent of Netflix and others are viewed on a TV, that's a total of four hours and eleven minutes of video content viewed on TVs per day by the average UK adult.

All the data on in-use and standby power consumption of the CRT, LED-backlit LCD, and plasma TV come from Ireland's Electricity Supply Board (2009) and for the 55-inch LED TVs from LG, https://tinyurl.com/lg-55-led-TV.

For calculating the carbon footprint, power consumption per hour was multiplied by my own figure for the carbon intensity of US electricity with 0.65 kg CO_2e per kWh in 2019 (see A *unit of electricity*, p. 55). When allocating a share of standby to per-hour figures, I assumed a standby power consumption of 0.32 watts for the 13-inch MacBook Pro, 0.5 watts for the 55-inch LED TV, and 3 watts for the older TVs.

Embodied carbon estimates for the 32-inch LED-backlit LCD come from Huulgaard, R.D., Dalgaard, R., and Merciai, S. (2013), "Ecodesign requirements for televisions—is energy consumption in the use phase the only relevant requirement?" *International Journal of Life Cycle Assessment*, 18(5), 1098–1105, https://tinyurl.com/huulgaard2013.

For the 28-inch CRT screen, the embodied carbon is based on estimates for a 25-inch CRT TV by Feng, C., and Ma, X.Q. (2009), "The energy consumption and environmental impacts of a color TV set in China." *Journal of Cleaner Production*, 17(1), 13–25, https://tinyurl.com/feng-ma2009.

The embodied carbon footprint of the 55-inch LED TV comes from the report "Lean ICT" (2018) by the French think tank The Shift Project, based on Samsung manufacturer's data: The Shift Project (2019), "Lean ICT: Towards digital sobriety," https://tinyurl.com/shiftproject2019.

The figure for the 42-inch plasma TV comes from Stobbe, L. (2007), "EuP preparatory studies 'televisions' (lot 5): Final report," Fraunhofer IZM, https://tinyurl.com/stobbe2007.

For the 13-inch MacBook Pro, the embodied carbon estimates come from Apple's Life Cycle Analysis report for the 1.4 GHz Quad-Core processor with 128 GB storage model introduced in July 2019: https://tinyurl.com/apple-13-macbook-pro.

For electricity consumption, I used a US electricity intensity of 0.65 kg CO_2e per kWh and information on power use from Apple, https://tinyurl.com/macbook-13-specs.

I have adjusted the embodied emissions of TVs and MacBooks upward for the 40 percent truncation error that's common for the manufacture of IT, based on my own research. The carbon footprint of transmission of TV content comes from a recent study by Schien et al.: Schien, D., Shabajee, P., Chandaria, J., Williams, D., and Preist, C. (2020), "BBC R&D White Paper 372: Using behavioural data to assess the environmental impact of electricity consumption of alternate television service distribution platforms," https://tinyurl.com/schien2020.

11 I'm assuming that you keep your TV for four years. If you keep your TV longer, this would spread out the total embodied carbon footprint of 735 kg more.

12 Schien et al. (2020). See note 10, above.

This study of the carbon footprint of the BBC, based on 2016 data, includes use-phase emissions from playout and coding and multiplexing, networks, satellites, data center storage, access network equipment in the home, and the viewing device that is typical for each transmission method. Their estimates exclude the embodied emissions in the network and data center infrastructure. They estimate 0.07 kWh for digital terrestrial TV, 0.17 kWh for satellite TV, and 0.18 kWh for cable TV and BBC iPlayer. I have used a US electricity factor of 0.65 kg CO_2e per kWh and subtracted their estimate for the share of the viewing device to arrive at an estimate of emissions per hour for the transmission alone.

13 According to Apple, their TV 4K and Siri Remote 32 GB set-top box model has an embodied footprint of 40 kg CO_2e (including manufacture and transport), https://tinyurl.com/apple-stb.

Adjusted for a truncation error of 40 percent, that's 67 kg CO_2e. Apple assume a lifetime of four years, so that's 16.7 kg CO_2e per year and 11 g per hour if I assume an average use per day of four hours and eleven minutes.

14 The British Academy of Film and Television Arts' Albert Project estimates the average carbon footprint of an hour's worth of TV content was 13.5 tons of CO_2e in 2017 (Albert Annual Report 2018; note that this does not include other greenhouse gases and is not per person but total emissions independent of the number of viewers). This includes electricity used in studios, heating of offices, transport, accommodation, catering, materials used on set, and waste. The footprint is highest for dramas and lowest for sports, and across all program types, people transport plays the biggest role. For 2018, BAFTA estimated the average footprint of one hour's TV content was 9.2 tons CO_2 (personal communication with William Bourns at BAFTA).

15 The 13-inch MacBook Pro 1.4 GHz Quad-Core processor with 128 GB storage has an embodied footprint of 326 kg CO_2e compared to 735 kg for the 55-inch LED; its electricity consumption per hour has a footprint of only 2 g compared to 30 g for the 55-inch LED, assuming the laptop is used five hours a day and the TV the average four hours and eleven minutes. This is one of the lowest-carbon laptops; models of the same line but with bigger internal storage use slightly more carbon and larger laptops like the 16-inch MacBook Pro 2.3 GHz 8-core processor with 1 TB storage have an embodied footprint twice as high at 620 kg CO_2e and a much higher hourly carbon footprint for electricity of 14 g, https://tinyurl.com/apple-16-macbook-pro.

16 Schien et al. (2020). See notes 10 and 12, above.

17 I assume that a Wi-Fi router has about the same embodied footprint as a set-top box, but that you share it with one other person, so it's 5 g CO_2e per hour.

18 I spoke to several experts to bottom out this entry:

Chris Preist, professor of Sustainability and Computer Systems at the University of Bristol, believes that higher resolution images do not result in proportionally higher quantities of data being downloaded, because of the impact of compression algorithms.

Jens Malmodin, senior specialist at Ericsson and an expert in the energy and carbon footprint of networks, thinks that 4G and 5G mobile networks are more efficient than fixed access networks (on which Wi-Fi relies), because their energy use scales with the amount of data being transferred. But from 10 MB per second (MBps) upward, fixed networks are more efficient than mobile data. Netflix recommends 3 MBps for SD, 5 MBps for HD, and 25 MBps for UHD, https://tinyurl.com/netflix-data.

19 Sandvine estimates that Netflix accounts for 12.6 percent of worldwide downstream volume: Sandvine (2019), "Netflix falls to second place in global internet traffic share as other streaming services grow," 12 September, https://tinyurl.com/sandvine-netflix.

According to Sandvine (2019), video as a whole accounts for 61 percent of downstream traffic and 22 percent of upstream traffic globally in 2018: "Global internet phenomena report," https://tinyurl.com/sandvine2019.

The popularity of video streaming can drive expansion of the underlying internet infrastructure, which can in turn enable more data-intensive activities and thus lead to further growth in infrastructure and thus emissions, according to Preist et al. (2016): Preist, C., Schien, D., and Blevis, E. (2016), "Understanding and mitigating the effects of device and cloud service design decisions on the environmental footprint of digital infrastructure." *Proceedings of the 2016 CHI Conference on Human Factors in Computing Systems*, 1324–1337, https://tinyurl.com/preist2016.

20 Schien et al. report that cable or satellite set-top boxes have a power consumption of 20 watts when in active standby and 10 to 15 watts in passive standby. Left on standby an average amount of nineteen hours and forty-nine minutes, that's 49 kg CO_2e per year for active standby or 25–37 kg CO_2e for passive standby. Wi-Fi routers for video streaming also have a high standby. They use around 10 watts and don't even have a standby. Over the course of one year, they use around 30 kg CO_2e in electricity from being on all the time: Schien et al. (2020). See notes 10 and 12, above.

21 Based on US electricity at 0.65 kg CO_2e and gas at 0.225 kg CO_2e per kWh. The gas figure is based on those supplied by BEIS (2019), https://tinyurl.com/beis-emission-factors-2019, and the US electricity figure comes from my own input-output model for electricity emissions for different countries. Both are adjusted to take account of power station supply chains and distribution. The average price per kWh is $0.13, according to https://tinyurl.com/USelectricitycost.

22 A couple of months ago, Claire Perry MP told radio listeners that the UK gets 32 percent of our energy from renewables. Actually, we get 32 percent of only our *electricity* from renewables. The difference in the two statements is huge. The latter translates to the UK sourcing something like 5 percent of our total energy consumption from renewables, which is currently the case. I wasn't the only one to call her out on this, but I never saw an effort made to correct this misinformation to those who had heard her original false statement. If we are ever going to get on top of climate change, we are going to need to ensure politicians get better at honoring the truth. Much more on this in one of my other books, *There Is No Planet B*.

23 Confederation of Paper Industries (2006), "UK paper making industries statistical fact sheet," www.paper.org.uk/info/reports/fact2006colour0707.pdf; Alsema, E.A. (2001), ICARUS 4: *Sector Study for the Paper and Board Industry and the Graphical Industry*, Utrecht Centre for Energy Research.

24 Based on figures for the carbon intensity of mixed paper to landfill from the UK government's Department for Business, Energy and Industrial

Strategy (BEIS) conversion factors for 2019: BEIS (2019), "Greenhouse gas reporting: conversion factors 2019," https://tinyurl.com/beis-emission-factors-2019.

25 For the mailing preference service (UK), you can do this online at https://tinyurl.com/mail-pref. For more information on opting out of junk mail, see https://www.consumer.ftc.gov/articles/how-stop-junk-mail for the US and https://www.canadapost-postescanada.ca/cpc/en/support/kb/receiving/mail-delivery/how-to-stop-receiving-advertising-mail for Canada.

26 It is already using more energy on Bitcoin mining than on powering all its households: Baraniuk, C. (2018), "Bitcoin energy use in Iceland set to overtake homes, says local firm," 12 February. BBC, https://tinyurl.com/baraniuk2018.

27 All the numbers on waste impacts come from the UK government's BEIS (2019) conversion factors. See note 24, above.

Conversion factors for virgin and recycled paper came from Confederation of Paper Industries (2006), "UK paper making industries statistical fact sheet." See note 23, above.

Environmental Defense Fund's report (1995), "Energy, air emissions, solid waste outputs, waterborne wastes and water use associated with component activities of three methods for managing newsprint," provided a sense of, and some figures for, transport and printing impacts. Added together, this gives a calculation of 1.5 kg CO_2e per kg for recycling a newspaper and 3.8 kg CO_2e per kg for sending it to landfill. It should also be noted that, when we weighed the weekly papers in the office, we found that they were a bit lighter than they were ten years ago, hence the slightly smaller footprint of each paper. I'm not entirely sure why this is, but it could come down to each individual sheet being slightly thinner and/or smaller, each complete paper containing fewer pages, there being fewer supplements, or a combination of the three.

28 Jern, M. (2018), "How many people consume bottled water globally?," 29 October. TAPP Water, https://tinyurl.com/jern2018.

29 The water itself is just 30 g CO_2e per 500 ml, according to bespoke work done by Small World Consulting with Booths Supermarkets: Berners-Lee, M., Moss, J., and Hoolahan, C. (2014), "The greenhouse gas footprint of Booths." Small World Consulting, https://tinyurl.com/booths-footprint.

The carbon intensity of PET, from which bottles are typically made, is 4.1 kg CO_2e per kilo (see *1 kg (2.2 lbs) of plastic*, p. 106, for more details). The bottles I weighed averaged around 50 g per quart of capacity, which is around 103 g CO_2e per 500 ml plastic bottle. Short-distance road transport has a footprint of 25 g CO_2e per bottle, whereas long-distance road transport is 101 g CO_2e.

30 In 2019, London installed fifty fountains around the city. This is a small start, at least. Greater London Authority (2019), "Mayor reveals locations of 50 new water fountains," 18 July, https://tinyurl.com/london-fountains.

31 In the EU, it is illegal to use the terms like milk, cheese, and yogurt for dairy alternatives: Beret, C. (2019), "Soy milk vs. EU law: Who's really harmed by labeling bans?," 22 March, *Medium*, https://tinyurl.com/beret2019.

32 Morris, H. (2019), "Global average per capita tissue consumption stands at above 5 kg—but 10 kg is possible," 1 December. *Tissue World Magazine*, https://tinyurl.com/world-tissue.

33 Roos, G. (2009), "Tesco to roll out carbon labels for TP and kitchen towels," 22 May. Environment + Energy Leader, https://tinyurl.com/roos2009. Tesco had a short-lived campaign of carbon labeling their products from 2009 to 2012. I'm still going to go with the numbers they used for toilet paper, as they're the only source that I can find on this.

34 For washing dishes by hand, I have used a 90 percent efficient boiler with a temperature of 50°C and 31.5 quarts of water—half of what the average person uses for doing a full load of dishes according to Stokes, T. (2006), "Washing dishes by hand wastes water," *EcoStreet*, 21 July, https://tinyurl.com/washing-dishes. At the higher end of hand washing, I have used a 60 percent efficient boiler at a scalding 65°C and using 126 quarts of water (twice the average).

35 According to the appliance manufacturer Gorenje, the bacterial counts per plate is 390 for hand washing compared to one in a dishwasher: Gorenje (2017), "What is more efficient: Washing the dishes by hand or using a dishwasher?," 24 March, https://tinyurl.com/gorenje-bacteria.

36 For the dishwasher, I have taken into account both the use-phase and the embodied emissions of the dishwasher itself. The environmentally extended input-output (EEIO) model gives us a figure of 270 kg CO_2e for a brand-new dishwasher, which works out as 0.1 kg CO_2e per wash assuming four washes a week over a twelve-and-a-half-year lifetime. For the use phase, I have used the consumption figures of a Bosch A+ efficiency model for the 65°C intensive wash and a 50°C economy wash and multiplied the kWh by the UK emissions factor, Bosch, "Series 6 free-standing dishwasher 60 cm white," https://tinyurl.com/dishwasher-bosch.

37 A whole bottle of dishwashing liquid probably has a footprint of about 1 kg CO_2e.

38 For the number up to the farm gate, I've used a 2008 report by the UK government's Department for Environment, Food and Rural Affairs (Defra): Defra (2008), "Final report for Defra project FO0103: Comparative life cycle assessment of food commodities procured for UK consumption through a diversity of supply chains," https://tinyurl.com/defra-fruit.

500 grams to 1 kilo (1.1 to 2.2 pounds)

1 The figures come from models used by Small World Consulting (www.sw-consulting.co.uk). An input-output approach is used for the fuel supply chains and the depreciation of the embodied emissions in the car and its manufacture.
2 Derived from the UK government's Department for Environment, Food and Rural Affairs (Defra) (2008), "Passenger transport emissions factors: methodology paper." Available from www.defra.gov.uk/environment/business/reporting/pdf/passenger-transport.pdf.
3 All mugs vary. This is based on a report by Starbucks on their own mugs. Can we trust a source with such obvious interests? Well, the results more or less square with my back-of-envelope calculations, http://business.edf.org/files/2014/03/starbucks-report-april2000.pdf.
4 I've used the numbers up to the farm gate from the following studies. For instant coffee, I've used Busser, S., Steiner, R., and Jungbluth, N. (2008), "LCA of packed food products: The function of flexible packaging," www.seeds4green.org/sites/default/files/ESU_-_Flexible_Packaging_2008_-_Exec_Sum.pdf. For ground coffee I've used Humbert, S., Loerincik, Y., Rossi, V., Margni, M., and Jolliet, O. (2009), "Life cycle assessment of spray dried soluble coffee and comparison with alternatives (drip filter and capsule espresso)." *Journal of Cleaner Production*, 17, 1351–1358, www.sciencedirect.com/science/article/pii/S0959652609001474. And for tea I've used Doublet, G., and Jungbluth, N. (2010), "Life cycle assessment of drinking Darjeeling tea: Conventional and organic Darjeeling tea," http://docplayer.net/31742051-Life-cycle-assessment-of-drinking-darjeeling-tea.html. For all post-farm-gate processing, I've used data from work with Booths.
5 All my calculations are based on a 250 ml mug. I've allowed 1.5 g of tea (with a footprint of 6.2 kg CO_2e per kg), 2 g of instant coffee (with a footprint of 17.5 kg CO_2e per kg), and 9 g of ground coffee (with a footprint of 5.8 kg CO_2e per kg). I've assumed 25 ml of added milk (excluding the lattes).
6 The data I've used comes from the Environmental Protection Agency (EPA) website: https://tinyurl.com/USwastestats.
7 The breakdown of emissions from bread up to the farm gate comes from Goucher, L., Bruce, R., Cameron, D., Koh, S.C.L., and Horton, P. (2017), "The environmental impact of fertilizer embodied in a wheat-to-bread supply chain." *Nature Plants*, 3, 1–5, www.nature.com/articles/nplants201712#Sec7.
8 From Stewart, T. (2009), *Waste: Uncovering the Global Food Scandal* (New York: W.W. Norton), which can be found here: https://tinyurl.com/TristramStuartWaste.
9 This comes from a piece of Small World Consulting work from 2015, "The Greenhouse Gas Footprint of Everards Brewery Ltd."
10 I have based the pizza base and tomato sauce on a BBC food recipe: https://tinyurl.com/bbc-pizza. The pizza base is made with 150 g of flour plus

some yeast and vegetable oil, and the tomato base is 100 ml of tomato paste with some herbs, garlic, and salt. Each pizza also has a fresh tomato on top, which I'm assuming are the loose salad variety, which have been grown somewhere in the South of Europe and transported a few thousand miles by road.

11 For mackerel, cod, tuna, and trout, I have used numbers up to the farm gate from Nielsen, P.H., Nielsen, A.M., Weidman, B.P., Dalgaard, R., and Halberg, N. (2003), "LCA food data base: Lifecycle assessment of basic food," (2000 to 2003). Aarhus University, Denmark. For shrimp and lobster, I have used numbers from Poore and Nemecek (2018), "Reducing food's environmental impacts through producers and consumers." *Science*, 992(6392), 987–992, https://tinyurl.com/poore2018. For post-farm-gate processing, I have used data from my work with Booths: Berners-Lee, M., Moss, J., and Hoolahan, C. (2014), "The greenhouse gas footprint of Booths." Small World Consulting, https://tinyurl.com/booths-footprint.

12 Not to plug my other books too much, but I do go into more detail in *There Is No Planet B* (2019).

13 A 2014 study estimated that restoring 424,000 hectares of rainforest in Brazil would cost $198 million per year for three years, followed by a drop in price thereafter, so I'm going to assume a total price tag of $972 million. I've assumed each hectare will sequester 500 tons CO_2e (assuming it's left alone for at least 200 years), so despite the hefty cost it would still work out at a massive 220 tons CO_2e per £1 ($2–3) spent, www.sciencemag.org/news/2014/08/affordable-price-tag-saving-brazils-atlantic-rainforest.

14 Based on $8,760 per panel in the US and a carbon saving of 16,800 kg CO_2e over the lifetime of the panel.

15 This figure is drawn directly from work I recently did for an app developer, part of which looked at the emissions of the average UK grocery basket. Converted to US dollars for this book.

16 My gasoline figure is based on a gasoline cost of 22 cents per mile for an average gas-powered car. This takes account of the extraction, shipping, and refining of the fuel, but not the depreciation or maintenance of the car.

17 The average price per kWh is 13 cents, according to https://tinyurl.com/USelectricitycost, so 4.97 kg CO_2e per $1 spent for US electricity (0.65 kg CO_2e/kWh).

18 "Sustainable lifestyle" is a tricky expression. It doesn't bear much scrutiny and we could get hopelessly bogged down defining it. However, I strongly suspect that whatever your definition, I would still stand by my assertion that it leaves plenty of scope for reading.

19 I'm using conversion factors of 1.59 kg, 2.59 kg, and 2.70 kg CO_2e per kilo for printing on recycled, typical UK mix, and 100 percent virgin paper. Confederation of Paper Industries (2006), "UK paper making industries statistical fact sheet," www.paper.org.uk/info/reports/fact2006colour0707.pdf.

20 A life-cycle analysis by Dowd-Hinkle (2012) estimated the footprint of an Amazon Kindle at 22.3 kg CO_2e, of which 21.5 kg CO_2e are from the manufacturing and transport. I have adjusted the latter figure upward to adjust for the truncation error that most life-cycle analyses make when estimating embodied footprint and which is about 40 percent of the "real" embodied footprint. Dowd-Hinkle estimates the use-phase electricity at only 0.6 kg CO_2e over three years, Dowd-Hinkle, D.J. (2012), "Kindle vs. printed book: An environmental analysis." Thesis, Rochester Institute of Technology, https://tinyurl.com/dowd-hinkle2012. I come to a similar figure when I use the approach I took in the last edition of this book. At the time of writing, you can get an Amazon Kindle Paperwhite (released 2018) for £119.99 ($169.15 US). If I multiply this by the emissions per pound spent in the computer manufacturing sector—0.180 kg CO_2e per £1—I get to 22 kg CO_2e. If we compare this to the embodied footprint of an iPad, which comes in at 130 kg CO_2e (217 kg if adjusted for truncation error), the Kindle has a much lower footprint. Based on an Apple LCA for the 12.9-inch iPad Pro (fourth generation) with 128 GB: https://tinyurl.com/apple-ipad.

21 This comes from working out the carbon footprint of a kWh of heat and a kWh of electricity in the US. A typical bath can hold about 32 gallons (with you in it, too). I've taken the cold-water temperature to be 10°C and a comfortable bath temperature to be 40°C. The heat capacity of water is 4.2 kilojoules per quart. I've assumed a 90 percent efficient boiler and used a conversion factor of 0.221 kg CO_2e per kWh for heat produced by natural gas (this uses a figure from Defra for the direct emissions of burning gas and adds to that a figure from our input-output model to estimate the supply chain impacts). There are 3,600 kilojoules per kWh. The footprint of the bath in kg CO_2e is $120 \times (39 - 8) \times 4.2 \times 0.225/(3{,}600 \times 90\%)$. For US electricity, I've used the same formula described above but with a carbon footprint of 0.65 kg CO_2 per kWh generated (again, this is the figure from Defra plus the figure from the input-output model to add on the estimate of supply chain impacts). The footprint of the water consumption is negligible.

1 to 10 kilos (2.2 to 22 pounds)

1 Emissions up to the farm gate are from Williams, A.G., Audsley, E., and Sandars, D.L. (2006), "Determining the environmental burdens and resource use in the production of agricultural and horticultural commodities." Defra (Department for Environment, Food and Rural Affairs) Research Project IS0205, Cranfield University and Defra, https://tinyurl.com/williams2006.

Emissions beyond the farm gate are based on a study by Small World Consulting for Booths Supermarkets: Berners-Lee, M., Moss, J., and

Hoolahan, C. (2014), "The greenhouse gas footprint of Booths." Small World Consulting, https://tinyurl.com/booths-footprint.

2 The figure for soy milk and the higher figure for global average milk can be found in Poore, J., and Nemecek, T. (2018). "Reducing food's environmental impacts through producers and consumers," *Science*, 992(6392), 987–992, https://tinyurl.com/poore2018. Thanks also to Joseph Poore for kindly providing additional data for oat, rice, and almond milk.

3 Numbers up to the farm gate come from Garnett, T. (2007), "The alcohol we drink and its contribution to UK greenhouse gas emissions—a discussion paper," Food Climate Research Network, https://tinyurl.com/garnett2007.

The rest of my numbers come from my work with Booths Supermarkets: Berners-Lee et al. (2014). See note 1, above.

4 For this I've used the thesis of a master's student I co-supervised: Swinn, R. (2017), "A comparative LCA of the carbon footprint of cut-flowers: British, Dutch and Kenyan. Environment and development." Thesis, Lancaster University. The thesis is summarized here: https://tinyurl.com/footprint-flowers.

5 Impacts up to the farm gate are from Williams et al. (2006). See note 1, above.

Impacts from the farm to the checkout are from Small World Consulting's work for Booths Supermarkets: Berners-Lee et al. (2014). See note 1, above.

6 I've taken the data from Poore and Nemecek (2018). See note 2, above. In almost all cases, the results are strikingly similar to my own research, including my work for supermarkets—unsurprisingly, because our sources are mainly the same.

7 See, for example, Lappeenranta-Lahti University of Technology (2017), "Protein produced from electricity to alleviate world hunger," https://tinyurl.com/protein-electricity. Thanks also to Channel 4 and George Monbiot for an interesting piece on this in their 2020 documentary *Apocalypse Cow: How Meat Killed the Planet*, https://tinyurl.com/apocalypse-cow.

8 Article reporting Singapore's trial of lab-grown meat: https://tinyurl.com/Singaporelabgrownmeat.

9 I've based all my numbers on the consumption figures of Bosch A+++ Series 4 washing machines (and Series 6 tumble dryers), https://tinyurl.com/washing-machine-bosch. The Small World Consulting (SWC) environmentally extended input-output (EEIO) model has an estimated carbon footprint of 281 kg CO_2e for the washing machine, and I have assumed that you'll get through two loads a week for a ten-year lifetime.

10 The rating system seems a bit unfair on condensing dryers since it doesn't take account of the fact that they keep the heat in the home instead of belching it into the outside world.

11 Reverse osmosis (RO) desalination of seawater produces anywhere between 0.4 and 6.7 kg CO_2e per cubic meter, compared to 0.3–34.3 kg CO_2e per cubic meter for thermal desalination techniques, according to a literature review by Cornejo, P. K., Santana, M.V., Hokanson, D.R., Mihelcic, J.R., and Zhang, Q. (2014), "Carbon footprint of water reuse and desalination: A review of greenhouse gas emissions and estimation tools." *Journal of Water Reuse and Desalination*, 4(4), 238–252, https://tinyurl.com/cornejo2014. Reverse osmosis desalination of brackish water is lower carbon (0.4–2.5 kg CO_2e per cubic meter) and water reuse lower still (0.1–2.4 kg CO_2e per cubic meter).

12 Global desalination was at 95 million cubic meters per day in 2019 (up from the 2009 estimate of 60 million in the first edition of this book) according to Jones, E., Qadir, M., van Vliet, M.T., Smakhtin, V., and Kang, S.M. (2019), "The state of desalination and brine production: A global outlook." *Science of the Total Environment*, 657, 1343–1356, https://tinyurl.com/jones-2019. Forty-eight percent of this desalination takes place in the Middle East and North Africa regions, and 55 percent of brine production takes place in these regions (brine production is at 141.5 million per cubic meter per day).

Another study comes out with similar numbers and the low-end estimate of global annual emissions from desalination: Duong, H.C., Ansari, A.J., Nghiem, L.D., Pham, T.M., and Pham, T.D. (2018), "Low carbon desalination by innovative membrane materials and processes." *Current Pollution Reports*, 4(4), 251–264, https://tinyurl.com/duong-2018.

13 Readers who recall the first edition of *Bananas* may notice that this is a lot less than the 1.6 percent of global annual emissions I estimated desalination as responsible for ten years ago. I overestimated my calculations a bit in the old book and 0.15–0.4 percent of global emissions is much more realistic.

14 Information on Seawater Greenhouse Ltd. comes from Wikipedia, https://tinyurl.com/wiki-seawater.

15 Chen, W., Chen, S., Liang, T., Zhang, Q., Fan, Z., Yin, H., et al. (2018), "High-flux water desalination with interfacial salt sieving effect in nanoporous carbon composite membranes." *Nature Nanotechnology*, 13(4), 345–350, https://tinyurl.com/chen-2018. See also Wang, H. (2018), "Low-energy desalination." *Nature Nanotechnology*, 13(4), 273–274, www.nature.com/articles/s41565-018-0118-y.pdf.

16 US Department of Agriculture (USDA) Food Safety and Inspection Service (2013), "Molds on food: Are they dangerous?," https://tinyurl.com/usda2013.

17 Williams et al. (2006). See note 1, above.

18 Williams et al. (2006). See note 1, above.

19 International Rice Research Institute, www.irri.org/.

20 There were 518 billion tons of rice produced in 2016: Grain Central (2018), "Global rice consumption continues to grow," 26 March, https://tinyurl.com/grain-central. I imagine that the global retail emissions will be much smaller than they are in the UK, so I've gone for the lower estimate to calculate total emissions of around 3.5 percent.

21 All statistics come from the International Rice Research Institute's *Rice Stats* database, https://tinyurl.com/rice-stats.

22 All the data for plastic, with the exception of nylon, has come from the following study: Zheng, J., and Suh, S. (2019), "Strategies to reduce the global carbon footprint of plastics." *Nature Climate Change*, 9(5), 374–378, https://tinyurl.com/zheng2019.

23 The number for nylon comes from Hammond, G.P., and Jones, C.I. (2019), *Inventory of Carbon and Energy* (ICE) *Database* V3.0. "Embodied carbon—the ICE database," circularecology.com. Available for download from https://tinyurl.com/ice-database.

24 This report highlights some of the health problems: Azoulay, D., Villa, P., Arellano, Y., Gordon, M.F., Moon, D., Miller, K.A., and Thompson, K. (2019), "Plastic and health: The hidden costs of a plastic planet," Center for International Environmental Law, https://tinyurl.com/azoulay2019.

25 Geye, R., Jambeck, J., and Law, K.L. (2017), "Production, use, and fate of all plastics ever made." *Science*, 3(7), 1–5, https://tinyurl.com/geyer2017.

26 Department for Environment, Food and Rural Affairs (Defra) (2008), "Final report for Defra project FO0103: Comparative life cycle assessment of food commodities procured for UK consumption through a diversity of supply chains," https://tinyurl.com/defra-fruit.

The two lower numbers here come from Williams et al. (2006). See note 1, above.

27 Williams et al. (2006). See note 1, above.

Although this looks like the best information around, it is contested. I know farmers who are highly critical of the assumptions made in the same report about organic dairy herds. My high-end figure is adjusted upward from Cranfield's 38.6 kg CO_2e per kilo to take account of produce from a colder time of year, rather than the year-round average reported in the Cranfield study. Since the first edition of the book, I have updated the figure with up-to-date electricity numbers up to the farm gate to give a number of 54 kg CO_2e.

28 Williams et al. (2006). See note 1, above.

Defra (2008). See note 26, above.

10 to 100 kilos (22 to 220 pounds)

1 The widow of former Philippines president Ferdinand Marcos was listed by *Newsweek* as one of the 100 "greediest people of all time." She gained some of her notoriety from her shoe collection, gathered while plenty of her fellow citizens lived in poverty.

2 A weakness of the input-output model I used for this is that it assumes that Chinese production is as carbon efficient as UK manufacture. It isn't. It's worse. In reality, a key carbon decision for footwear suppliers is where to have products made.

3 For example, Periyasamy, A.P., and Durasisamy, G. (2019), "Carbon footprint on denim manufacturing." *Handbook of Eco-Materials*, 1581–1598, https://tinyurl.com/periyasamy2019.

4 Environmental Improvement Potential of Textiles (IMPRO-Textiles), JRC Scientific and Technical Reports: Beton, A., Dias, D., Farrant, L., Gibon, T., Le Guern, Y., Desaxce, M., et al. (2014), "Environmental improvement potential of textiles (IMPRO-textiles)," European Commission, https://tinyurl.com/beton2014.

5 Niinimäaki, K., Peters, G., Dahlbo, H., Perry, P., Rissanen, T., and Gwilt, A. (2020), "The environmental price of fast fashion." *Nature Reviews Earth and Environment*, 1, 189–200, https://tinyurl.com/niinimaki2020.

6 Clothing footprint of the UK, comparison to global estimates, based on House of Commons Environmental Audit Committee (2019), "Fixing fashion: Clothing consumption and sustainability," https://tinyurl.com/fixing-fashion.

7 This is the additional footprint arising from your decision to make the commute, given that everyone else is already on the road. It is also the difference you can make by stopping commuting. It is more than your fair share of the total pollution, which would only be double rather than three times the normal emissions from driving that distance on an empty road.

8 To make it very simple, think of a queue ten cars long, moving at one car per minute. Assuming the queue has stayed the same size, those ten cars will, between them, have queued for a hundred car minutes by the time they have all gone through. Add your car and you have eleven cars all queuing for eleven minutes. That's twenty-one minutes more queuing, even though you experience just eleven minutes. You get the same effect when you model slightly more complicated things such as ring roads with queues at each roundabout. None of this takes account of the possibility that you are the person who gets stuck at the junction, triggering gridlock and a whole new multiplier effect.

9 The Highway Code figures for typical stopping distances are 96 m (24 car lengths) at 70 mph and just 53 m (13 car lengths) at 50 mph, https://tinyurl.com/highway-code. The stopping distance has two components:

the thinking distance, which is proportional to your speed, and the larger braking distance, which is proportional to the square of your speed. On this basis, a lane at 50 mph can take nearly twice the traffic of one at 70 mph. So, there is no need for anyone to queue when the lane closes, provided that no one leaves it to the last moment to change lanes. In reality, most drivers don't leave as much as their stopping distance between them and the car in front, but the principles here still apply if they keep leaving the same proportion of that stopping distance between themselves and the next car as they slow down.

10 This information comes from a Chatham House report: Lehne, J., and Preston, F. (2018), "Making concrete change: Innovation in low-carbon cement and concrete," https://tinyurl.com/lehne2018.

11 Vizcaíno-Andrésa, L.M., Sánchez-Berriela, S., Damas-Carrerab, S., Pérez-Hernándezc, A., Scrivenerd, K.L, and Martirena-Hernández, J.F. (2015), "Industrial trial to produce a low clinker, low carbon cement." *Materiales de Construccion*, 65, 317, https://tinyurl.com/vizcaino2015.

12 Maddalena, R., Roberts, J., and Hamilton, A. (2018), "Can Portland cement be replaced by low-carbon alternative materials? A study on the thermal properties and carbon emissions of innovative cements." *Journal of Cleaner Production*, 186, 933–942, https://tinyurl.com/maddalena2018.

13 A staggering five hours of life are lost through death per 1,000 miles of driving. My sum was just this: loss of life expectancy per mile = 2,538 deaths on UK roads per year × 48 remaining years of life expectancy of an average driver, divided by 216 billion person-car miles on UK roads per year = 5 hours life lost per 1,000 miles of driving (National Travel Survey, Department for Transport, 2009). I've based my sums on your having a life expectancy of another forty-eight years (I picked a forty-year-old man with a healthy lifestyle because it gives me a nice round number), but you might want to adjust for your own situation. I haven't taken account of the fact that some of the deaths are of pedestrians (thinking that you might be just as bothered about killing others as you are yourself), but I also haven't taken into account the possibility that you might acquire one of the 26,000 serious injuries or 150,000 minor injuries that are served up to UK car users each year. It's a lot better to be injured than killed on the road, but injury happens ten times more often. I have also assumed that highway journeys are averagely safe per mile compared with other car trips.

14 Travel time estimate based on the AA Route Planner, https://tinyurl.com/aa-route.

15 SUV safety comes from Consumer Reports: https://tinyurl.com/SUVsafety.

16 An iPhone 11 with 128 GB has an embodied carbon footprint of 63 kg CO_2e (this figure includes production and transport) according to Apple,

https://tinyurl.com/iphone11-footprint. I have adjusted the embodied figure upward for truncation error to 105 kg CO_2e. Apple is quite optimistic with the assumption of three years of life, but according to Belkhir and Elmeligi, the average life of a smartphone is two years, so we have assumed that for all the headline figures: Belkhir, L., and Elmeligi, A. (2018), "Assessing ICT global emissions footprint: Trends to 2040 and recommendations." *Journal of Cleaner Production*, 177, 448–463, https://tinyurl.com/belkhir2018.

Apple reports that the iPhone 11 has a battery capacity of 11.91 Wh, https://tinyurl.com/iphone11-specs and ZDNet estimates that this lasts for 1.6 days, https://tinyurl.com/zdnet-iphone, so one day uses 7.44 Wh. Using a US electricity intensity of 0.65 kg CO_2e per kWh and considering charger efficiency is around 85 percent, https://tinyurl.com/charger-efficiency, that's close to 3 g per day in electricity consumption.

The company RescueTime (https://blog.rescuetime.com/screen-time-stats-2018/) analyzed data from 11,000 users of their app and concluded that people spend an average of three hours and fifteen minutes on their phones. MacKay, J. (2019), "Screen time stats 2019: Here's how much you use your phone during the workday," 21 March, https://tinyurl.com/rescue-time. Allocating the electricity per day to the three hours and fifteen minutes that you use your phone actively, the footprint from electricity per minute is 0.015 g CO_2e, or 1.1 kg CO_2e per year. When you use your phone just one hour per day, it's 0.3 kg CO_2e per year and, for using it ten hours a day, it's 3.3 kg CO_2e per year. Clearly, it doesn't make much of a difference because of the size of the embodied footprint.

For the use of networks and data centers, I have used estimates of electricity use in networks and data centers from Ericsson: 5 watts for the networks, 5 watts in data centers, and 10 watts for your Wi-Fi router, but this is split across two people, so 5 watts per user. The footprint of doing something online depends on how long you spend on your device and on the internet: Ericsson (2020), "A quick guide to your digital carbon footprint—deconstructing information and communication technology's carbon emissions," https://tinyurl.com/ericsson2020, and specifically their background report, www.ericsson.com/4906b7/assets/local/reports-papers/consumerlab/reports/2020/background-calculations-to-true-and-false-report.pdf.

I'm using a global electricity mix carbon intensity for the networks and data centers, because you might well use some that are in other countries when you use the internet, but for the router I have assumed UK electricity. I have assumed that you have a Wi-Fi router running twenty-four hours a day and allocated the electricity to the average two-and-a-half hours that people use the internet at home per day, according to Center for the Digital Future Report (2018), "The 2018 digital future report: Surveying the digital future," https://tinyurl.com/digitalfuture2018.

That comes to 0.4 g CO_2e per minute (22.5 kg per year) Wi-Fi use if you use the internet for three hours and fifteen minutes. Over the course of a year, that adds up to 15 kg, or 17 kg (or 42 kg), if you use your phone for one (or ten) hours, respectively; assuming you also use one other device to access the Wi-Fi, only half of the Wi-Fi router's electricity use per person belongs to the smartphone. The footprint is not directly proportional to how long you use it actively, because of the way that the Wi-Fi router is allocated. I have assumed that you use it for one hour or eight hours per day, respectively. In the case of using your phone for one or ten hours every day, I have allocated the Wi-Fi router's daily electricity use to one and eight hours per day, respectively. That comes to 17 kg per year for one hour of use per day, or 42 kg per year for ten hours per day, for the internet use. If you use mobile networks, which don't require a Wi-Fi router, it's just 0.1 g per minute.

Ericsson arrives at a slightly higher figure; they've calulated a year's use of the internet through your smartphone at a footprint of 25 kg CO_2e, based on a global electricity factor of 0.6 kg CO_2e per kWh. When I adjust this for my electricity figure of 0.63 kg CO_2e per kWh, I get 26.3 kg. Ericsson bases this on four hours of use per day. For three hours and fifteen minutes, this breaks down to 21.3 kg CO_2e per year, and 8.1 kg and 80.8 kg if you use the internet for one or ten hours per day, respectively.

17 Based on an iPhone 11, including 0.7 g for the embodied footprint, 0.2 g for use-phase footprint, and 0.4 g for electricity use in the network, data centers, and Wi-Fi router.

18 Belkhir and Elmeligi (2018). See note 16, above. This study concluded that the average lifetime of a smartphone is two years. This is not because the phone wouldn't last any longer but because phone contracts and fashion encourage people to change their phones more often than necessary. However, there is some evidence that the useful life of smartphones might be increasing, thereby spreading the embodied carbon footprint of the device over a longer time period. The NPD (2018) reported that, in the US, the average use has increased to thirty-two months in 2017, up from twenty-five months in 2016: Cited in Benzinga (2018), "The average upgrade cycle of a smartphone in the US is 32 months, according to NPD connected intelligence," 12 July, https://www.benzinga.com/pressreleases/18/07/p12009755/the-average-upgrade-cycle-of-a-smartphone-in-the-u-s-is-32-months-acco.

19 Cisco (2020) estimates that there were 7.3 billion mobile phones in 2018 and that there will be 8.2 billion in 2025. Based on this, I extrapolated 7.7 billion in 2020. I have multiplied this by 76 kg CO_2e per phone, based on the iPhone 11 used for three hours and fifteen minutes per day and kept for two years, assuming global electricity rather than US electricity. This includes the emissions embodied in the device, the electricity to run the

device and the networks and data centers, and is 1 percent of 2018 global greenhouse gas emissions (see *The Cloud and the world's data centers*, p. 179). Of course, not all the world's mobile phones are iPhones and the average of three hours and fifteen minutes use per day for smartphones might be different for other mobile phones, so this is just a rough estimate and probably an overestimate: Cisco (2020), *Cisco Annual Internet Report* (2018–2023) *White Paper*, https://tinyurl.com/cisco-report-2020.

20 Jens Malmodin, senior specialist at Ericsson and an expert in the energy and carbon footprint of networks, told me that he estimates that mobile networks used for transmitting texts and voice calls use on average 2 watts. For a US call using 0.65 kg CO_2e per kWh, a one-minute phone call would have a carbon footprint of 0.1 g CO_2e.

21 A European Environmental Bureau (EEB) report estimates that extending the lifetime of all smartphones by one year (three years, five years) would save 2.1 million tons CO_2e (4.3 million tons CO_2e, 5.5 million tons CO_2e) per year in the EU by 2030: EEB (2019), "Coolproducts don't cost the Earth," https://tinyurl.com/eeb-2019.

22 TechRepublic: https://www.techrepublic.com/article/10-places-to-recycle-your-cellphone/, HowStuffWorks: https://electronics.howstuffworks.com/donate-cell-phone-charity.htm, Consumer Reports: https://www.consumerreports.org/recycling/how-to-recycle-electronics/, or simply Google "recycle my cell."

100 to 1,000 kilos (220 pounds to 1 ton)

1 According to the US Energy Information Administration, https://www.eia.gov/tools/faqs/faq.php?id=23&t=10.

2 I have calculated the average carbon intensity of the economy for all countries classified as low income by the World Bank, using their estimates of consumption-based CO_2e emissions in 2016, where this data was available. Ritchie, H., and Roser, M. (2019), "CO_2 and greenhouse gas emissions," Our World in Data, https://tinyurl.com/ritchie2019—and dividing it by total GDP in 2016 provided by the World Bank, https://tinyurl.com/world-bank-gdp. The average carbon intensity is 0.64 kg CO_2e per US dollar GDP for low-income countries and 0.21 kg for the UK. So, on average, low-income countries have a carbon intensity that is 3.1 times higher than that of the UK. This is a very crude measure, of course.

3 Phillips, D. (2018), "Illegal mining in Amazon rainforest has become an 'epidemic,'" *Guardian*, 10 December, https://tinyurl.com/phillips-2018.

According to research by the Amazon Geo-Referenced Socio-Environmental Information Network (RAISG), using satellite images of the Amazon, there has been an exponential increase in illegal gold mining in recent years, including in Indigenous and protected natural areas. This has been partly sparked by spikes in gold prices after the 2009 recession.

The mining does not just lead to deforestation, but also mercury contamination, which is used for purifying the gold and has ill effects on ecosystems and human health.

4 Some of these have been listed in this article by the *Independent*: Bergman, S. (2019), "9 best ethical and sustainable jewelry brands." *Independent*, 5 February, https://tinyurl.com/bergman2019.

The Fairtrade Foundation also lists some: Hackett, F. (2020), "7 ethical and sustainable jewelry brands," 12 February, https://tinyurl.com/hackett2020.

5 The figure of 20 percent of presents being unwanted comes from: Haq, G., Owen. A., Dawkins, E., and Barrett, J. (2007), "The carbon cost of Christmas." Stockholm Environment Institute, https://tinyurl.com/haq2007.

6 Taken from the ICE database: Hammond, G.P., and Jones, C.I. (2019), *Inventory of Carbon and Energy* (ICE) *Database* V3.0, circularecology.com. Available for download here from https://tinyurl.com/ice-database.

7 See note 6 above.

8 All information comes from the Energy Saving Trust (EST)'s website: https://tinyurl.com/est-loft.

9 The ICE database gives a figure of 1.28 kg CO_2e per kilo for wool. I've gone with this, despite the problems that process life-cycle analysis has with underestimating absolute numbers. Hammond and Jones (2019). See note 6, above.

10 Visit Energy.gov for information on federal energy efficiency tax credits and rebates, https://www.energy.gov/energysaver/services/incentives-and-financing-energy-efficient-homes. Many state and local governments offer additional support for improving the energy efficiency of your principal residence.

11 For embodied carbon, I used the following environmental reports: Apple MacBook Pro 13-inch, 1.4 GHz Quad-Core processor, 128 GB storage, introduced July 2019, https://tinyurl.com/apple-13-macbook-pro; Apple MacBook Pro 16-inch, 2.3 GHz 8-core processor, 1 TB storage, introduced November 2019, https://tinyurl.com/apple-16-macbook-pro; Dell Precision 5530, released June 2018, https://tinyurl.com/dell-5530; HP Chromebook 14 G5, introduced February 2018, https://tinyurl.com/hp-chromebook. HP does not give a percentage for the share of embodied carbon in the total footprint, so I have assumed the same percentage as for the Dell Precision 5530: 84 percent.

12 For electricity consumption, I used a US electricity intensity of 0.65 kg CO_2e per kWh and information on power use. For the Apple MacBooks, I divided the battery capacity by the length of time it lasts to arrive at a per-hour electricity consumption. This is based on data from Apple: https://tinyurl.com/macbook-13-specs.

Estimates for the HP laptop are based on its typical annual energy consumption of 19.7 kWh, assuming it is used three hours per day.

The other estimates for average laptops, desktop PCs, and gaming PCs are based on: Ericsson (2020), "A quick guide to your digital carbon footprint—deconstructing information and communication technology's carbon emissions," https://tinyurl.com/ericsson2020. Ericsson reports that an average laptop uses 30 watts, a PC with an external screen uses 150 watts, and a gaming PC with external screen uses 200 watts.

13 For the use of networks and data centers, I have used estimates of electricity use in networks and data centers from Ericsson: 5 watts for the networks, 5 watts in data centers, and 10 watts for your Wi-Fi router, but I'm assuming you share the Wi-Fi with one other person, so 5 watts per user. Ericsson (2020). See note 12, above, and specifically their background report, https://tinyurl.com/ericsson2020-background.

I'm using a global electricity mix carbon intensity for the networks and data centers, because you might well use some that are in other countries when you use the internet, but for the router I have assumed UK electricity. I have assumed that you have a Wi-Fi router running twenty-four hours a day, and allocated the electricity to the average two-and-a-half hours that people use the internet at home per day, according to Center for the Digital Future (2018), "The 2018 digital future project: Surveying the digital future," https://tinyurl.com/digitalfuture2018.

That comes to 22 g CO_2e per hour Wi-Fi use if you use the internet for three hours and fifteen minutes.

14 At Small World Consulting, we looked into some fairly standard methodologies used for life-cycle analyses of different types of products, including those used by Apple, and used this to map the likely truncations against a system-complete input-output methodology, to arrive at a figure for the proportion of all carbon that would be likely to be missed from the footprint. For different types of products, the proportion that we can expect to be truncated varies. For IT equipment, we came to a best estimate of 40 percent truncation error for the embodied carbon and 18 percent for the use phase. If we get time, we'll publish what we did!

15 De Decker, K. (2009), "The monster footprint of digital technology," *Low-Tech Magazine*, 16 June, https://tinyurl.com/dedecker2009. I'm guessing that the carbon footprint per gram is the same as it was ten years ago, when I wrote the first edition of this book, but of course a 2 g chip is a lot more powerful these days than it was back then. The paper this article refers to is Williams, E., Ayres, R., and Heller, M. (2003), "The 1.7 kilogram microchip: Energy and material use in the production of semiconductor devices." *Environmental Science and Technology*, 36(24), 5504–5510, www.researchgate.net/publication/5593533_The_17_Kilogram_Microchip_

Energy_and_Material_Use_in_the_Production_of_Semiconductor_Devices.

16 For the goldfish, I've assumed a 45-quart tank containing two goldfish. I've estimated that the tank will use up about 150 kWh per year, which works out at roughly 25 kg CO_2e per year: Algone, "Aquarium power consumption. Energy cost of a fish tank," https://tinyurl.com/fishtank-energy. I've assumed the food is negligible.

17 The study I've based the numbers on comes from Martens, P., Su, B., and Deblomme, S. (2019), "The ecological paw print of companion dogs and cats." *BioScience*, 69(6), 467–474, https://tinyurl.com/martens2019.

18 See, for example, Henriques, J. (2020), "Can dogs be vegetarian?" *Dogs Naturally*, 6 March, https://tinyurl.com/henriques2020.

19 Su, B., Martens, P., and Enders-Slegers, M.J. (2018), "A neglected predictor of environmental damage: The ecological paw print and carbon emissions of food consumption by companion dogs and cats in China." *Journal of Cleaner Production*, 194, 1–11, https://tinyurl.com/martens-2018.

20 According to Moneysavingexpert.com, the average cost of surgery at the vet is £1,500 (over $2,000): Schraer, N. (2018), "Millions of pet owners risk paying thousands of pounds in vet bills," 29 May, https://tinyurl.com/schraer2018. The environmentally extended input-output (EEIO) model has veterinary services at 0.144 kg CO_2e per £1 spent, giving the figure of 215 kg CO_2e.

21 Okin, G.S. (2017), "Environmental impacts of food consumption by dogs and cats." PLoS ONE, 12(8), 1–14, https://doi.org/10.1371/journal.pone.0181301.

22 Schwartz, L. (2014), "The surprisingly large carbon paw print of your beloved pet." *Salon*, 20 November, https://tinyurl.com/schwartz2014.

23 Small World Consulting did a piece of work for the Ecology Building Society. We calculated that their total emissions were 284 tons CO_2e in 2018, while their total income that year was £5,756,000. Both are reported in their 2019 annual report (https://tinyurl.com/ecologybsocwww.ecology.co.uk/wp-content/uploads/2020/04/EBS-Annual-Report-Accounts-2019.pdf). That means a carbon intensity of 49 grams per £1 ($1.50).

24 Oil Change International and several other campaign groups score major banks on their policies and practices around financing of fossil fuels: Oil Change International (2020), "Banking on climate change 2020: Fossil fuel finance report card," https://tinyurl.com/oilchangeint2020. JPMorgan Chase, Wells Fargo, Citi, and Bank of America finance fossil fuel the most. Several British banks, including Barclays, HSBC, Santander, and RBS/NatWest, have significant investments in fossil fuel, too. The international campaign group Fossil Banks No Thanks rates banks on their fossil-fuel investments, www.fossilbanks.org. US-based Mighty Deposits rates

American banks and credit unions on their environmental friendliness, including commitments not to invest in fossil fuels, https://mighty deposits.com/posts/environmentally-friendly-banks.

1 to 10 tons

1 Based on my input-output model.
2 Based on the average coronary bypass surgery cost. The NHS reports costs per intervention for 2018–19 on their website: https://tinyurl.com/nhs-costs. Prices vary between £9,888 (for an uncomplicated standard coronary artery bypass graft) and £17,977 (for a complex coronary artery bypass graft with single heart valve replacement or repair), with an average of £13,107. The average cost of a coronary bypass in the US in $40,000, https://www.healthgrades.com/right-care/tests-and-procedures/the-10-most-common-surgeries-in-the-u-s but can rise to over $75,000, https://blog.transonic.com/cardiothoracic-surgery/u.s.-coronary-artery-bypass-surgery-cost.
3 In 2018–19, 15,000 heart bypass operations, 23,000 major hip surgeries, and over 45,000 major knee surgeries were performed, according to the NHS. The average price of a hip surgery is £6,279, while knee surgeries each cost £6,108, on average. While these figures translate to $8,852 and $8,610 US, respectively, they do not represent the actual cost of these surgeries in the US, where joint replacement surgeries overall range from $16,500 to $33,000. Assuming they are as carbon intensive as the healthcare sector is on average, that would mean a carbon footprint of 1,108 kg and 1,078 kg CO_2e, respectively. Costs are drawn from the 2018–19 National schedule of NHS Costs (https://improvement.nhs.uk/resources/national-cost-collection/).
4 This includes medical and dental services in hospitals and practices, as well as the manufacture of pharmaceuticals, diagnostic tests, and wound dressings, and is based on my input-output model.
5 Based on a 2019 research paper: Pichler, P.-P., Jaccard, I., Weisz, U., and Weisz, H., "International comparison of health care carbon footprints," *Environmental Research Letters*, 14 (2019), 064004, https://tinyurl.com/healthcarefootprints.
6 Material Economics (2019), "Industrial transformation 2050—Pathways to net-zero emissions from EU heavy industry," https://tinyurl.com/materialecon2019.
7 Carbon Trust (2011), "International carbon flows: Steel, 2011," https://tinyurl.com/carbontrust2011. Read by eye from a diagram. At the time, about half of the steel for construction was in China, https://bit.ly/2Geuonj.
8 Thanks to Chris Goodall for this analysis: Goodall, C. (2020), "The extra costs of decarbonized steel." *Carbon Commentary*, 14 January, https://tinyurl.com/goodall2020, https://bit.ly/2sRallo.

9 Ammonium nitrate (NH_4NO_3) fertilizer is 35 percent nitrogen by weight. The nitrous oxide (N_2O) that is released is 64 percent nitrogen by weight. The 1 percent of the nitrogen that is emitted is 0.55 percent of the original weight of the fertilizer in N_2O, with a global warming potential 300 times that weight in CO_2 equivalent. So, 1–5 percent nitrogen released to the atmosphere is 1.65–8.25 tons CO_2e per ton of fertilizer applied to the crop. All the agricultural data in this section came from a lecture by Professor David Powlson during a visit to Lancaster University in November 2009. He is working with the Chinese government to get the message across to farmers.

10 China used 24 million tons of nitrogen fertilizer in 2017 (out of the world's 107 million tons annual total) according to the International Fertilizer Association, https://www.fertilizer.org//.

11 Cui, Z., Zhang, H., Chen, X., Zhang, C., Ma, W., Huang, C., et al. (2018), "Pursuing sustainable productivity with millions of smallholder farmers." *Nature*, 555(7696), 363–366, https://tinyurl.com/cui2018.

12 Based on UN data, Worldometer reports that China's population makes up 18.4 percent of the world population, as of May 2020, https://tinyurl.com/china-population, www.worldometers.info/world-population/china-population/.

According to the World Bank, China had 118,900,000 hectares of arable land, while the global number was 1,423,083,180 hectares, https://tinyurl.com/china-land. This compares with 24.9 percent arable land in the UK in 2016, for example.

13 Cui et al. report that a large-scale agricultural study giving advice to 20.9 million Chinese smallholder farmers between 2005 and 2015 led to 11 percent increased yields and 16 percent reduced fertilizer use, saving 1.2 million tons of nitrogen. The intervention reduced greenhouse gas emissions from fertilizer use by three-quarters: Cui et al. (2018). See note 11, above.

14 Here is a glimpse of the main issues. The amount of N_2O that a jet engine produces varies with altitude, and its effect on ozone levels also depends on altitude. Furthermore, the effect of that ozone on climate is altitude dependent. Planes also cause contrails under certain atmospheric conditions, and these are known to make a short-lived but large contribution to the greenhouse effect. The contrails themselves depend on temperature, weather conditions, time of day, and altitude.

15 Jones, C. (2020), "Circular ecology, embodied carbon of solar PV: Here's why it must be included in net zero carbon buildings." *Circular Ecology*, https://tinyurl.com/pv-footprint.

16 It assumes the world follows a 2°C pathway to 2050: Pehl, M., Arvesen, A., Humpenöder, F., Popp, A., Hertwich, E. G., and Luderer, G. (2017), "Understanding future emissions from low-carbon power systems by integration of life-cycle assessment and integrated energy modelling."

Nature Energy, 2 (12), 939–945. https://tinyurl.com/pehl2017, doi:10.1038/s41560-017-0032-9.

17 See this paper, which makes the case that the resources required to make renewables infrastructure are not a constraint: Diesendorf, M., and Wiedmann, T. (2020), "Implications of trends in energy return on energy invested (EROI) for transitioning to renewable electricity." *Ecological Economics*, 176: 106726, doi: 10.1016/j.ecolecon.2020.106726.

18 I'm using an emissions factor for natural gas of 240 g CO_2e per kWh. This is a little higher than the UK government's Department for Business, Energy and Industrial Strategy (BEIS) figures, as I've used input-output analysis to include something for the supply chain pathways that BEIS doesn't include. I'm taking the average cost of US solar panels from EnergySage: https://tinyurl.com/yb4px3fy.

10 to 1,000 tons

1 Prices are based on manufacturers' prices for the most basic and most high-end configurations. Citroën C1: https://tinyurl.com/citroen-prices; Ford Focus Titanium: https://tinyurl.com/ford-prices; Renault Zoe (electric): https://tinyurl.com/renault-prices; Toyota Prius Plug-in hybrid: https://tinyurl.com/toyota-prices; Range Rover Sport HSE: https://tinyurl.com/rangerover-prices.

2 For example, Carbon Brief estimates that going electric typically cuts lifetime emissions per mile to one-third of an average gasoline car, based on much lower estimates of the footprint of manufacture. The reason I trust my embodied carbon estimate more is that I suspect the process life-cycle analyses drawn upon here suffer from serious systematic underestimation, or truncation error (see *Where the numbers come from*, p. 232): Hausfather, Z. (2019), "Factcheck: How electric vehicles help to tackle climate change." Carbon Brief, 13 May, https://tinyurl.com/hausfather2019.

3 The emissions per fuel are based on rough numbers. In my calculations, I have assumed that kerosene releases 3,750 kg of greenhouse gases per ton, based on my input-output model factor for kerosene. Note that rockets use RP-1, a highly refined form of kerosene, so the exact emissions per ton will be slightly different.

The emissions per kilogram of liquid hydrogen, liquid oxygen, and hydrogen peroxide are based on the energy that is required to synthesize them and the carbon intensity of global average electricity. However, they exclude the emissions from transport and storage of the fuel and also exclude a share of the emissions embodied in the equipment needed to synthesize and transport the fuel. Electricity figures for hydrogen are based on 50 kWh per kg at 80 percent efficiency (according to Wikipedia, https://tinyurl.com/hydrogen-footprint), 0.23 kWh per kg for liquid oxygen (according to a report by the Gas Technology Institute, https://

tinyurl.com/liquid-oxygen-footprint), and 10 kWh per kg for hydrogen peroxide (based on a study by the Fraunhofer Institute, https://tinyurl.com/hydrogen-peroxide-footprint).

Based on my input-output model factor for kerosene, note again that rockets use RP-1, a highly refined form of kerosene, so the exact emissions per ton will be slightly different. As always, the carbon footprint numbers here are just an approximation.

4 According to Astronautix, kerosene consumption of 5,897 kg is based on information at the following link: https://tinyurl.com/new-shephard.

5 119,100 kg of kerosene based on the SpaceX Falcon 9 v1.1 (according to Spaceflight101, https://tinyurl.com/falcon-9).

6 According to Cool Cosmos, https://tinyurl.com/space-shuttle. I have assumed that the solid fuel has the same carbon intensity as kerosene.

7 All converted from quarts. Figures are from the website Space, https://tinyurl.com/moon-rocket. Shuttle data come from Wikipedia. Other figures in my calculations were: 31 MJ (1 megajoule equals 1 million joules) per kilo for the solid fuel; I used 0.07 kg CO_2e per MJ as a general figure for emissions from the burning of fossil fuels and added 10 percent for their supply chains up to the point of combustion; 143 MJ per kilo for the hydrogen; global average electricity intensity of 0.63 kg CO_2e per kWh, divided by 3.6 to get MJ >>0.175 kg CO_2e per MJ; 143 MJ per kilo of hydrogen; nothing added for supply chain; 25 kg CO_2e per kilo of hydrogen.

8 Richard Feynman's (1989) book *"What Do You Care What Other People Think?": Further Adventures of a Curious Character* (London: Unwin/Hyman) is a fascinating and entertaining account of the technical and management failures behind the disaster. It is also recommended for anyone who is trying to get some clear thinking into a bureaucracy.

9 Smoucha, E., Fitzpatrick, K., Buckingham, S., and Knox, O. (2016), "Life cycle analysis of the embodied carbon emissions from 14 wind turbines with rated powers between 50 kW and 3.4 Mw." *Journal of Fundamentals of Renewable Energy and Applications*, 6(4), 1–10, https://tinyurl.com/smoucha2016.

10 Hausfather, Z. (2018), "Analysis: How much 'carbon budget' is left to limit global warming to 1.5°C?" Carbon Brief, 9 April, https://tinyurl.com/hausfather2018 and https://bit.ly/3c0inhl. More details of carbon budgets in *There Is No Planet B* (2019).

Millions of tons

1 Aiuppa et al. estimate that Mount Etna releases around 2,000 tons CO_2 per day (that's 730,000 tons per year) during quiescent passive periods when the volcano is dormant: Aiuppa, A., Federico, C., Giudice, G., Gurrieri, S., Liuzzo, M., Shinohara, H., et al. (2006), "Rates of carbon dioxide plume degassing from Mount Etna volcano." *Journal of Geophysical Research: Solid Earth*, 111(B9), https://tinyurl.com/aiuppa2006.

The Holuhraun eruption in 2014 was the largest volcanic eruption in Iceland since 1783 and is estimated to have emitted 5.1 million tons CO_2: Pfeffer M.A., Bergsson, B., Barsotti, S., Stefánsdóttir, G., Galle, B., Arellano, S., et al. (2018), "Ground-based measurements of the 2014–2015 Holuhraun volcanic cloud (Iceland)." *Geosciences*, 8(1), 29, https://tinyurl.com/pfeffer2018.

Gerlach et al. estimate that the eruption of Mount Pinatubo in 1991 released 42 million tons CO_2: Gerlach, T.M., Westrich, H.R., and Symonds, R.B. (1996), "Preeruption vapor in magma of the climactic Mount Pinatubo eruption: Source of the giant stratospheric sulfur dioxide cloud." In Newhall, C.G., and Punongbayan, R.S. (eds.), *Fire and Mud: Eruptions and Lahars of Mount Pinatubo, Philippines* (Seattle and London: University of Washington Press), pp. 415–433, https://tinyurl.com/gerlach1996.

2 Research by the Deep Carbon Observatory program suggests that volcanic activity releases between 280 and 360 million tons CO_2 into the atmosphere and oceans every year. This includes active volcanic vents, released through soils, faults, and fractures in volcanic regions, volcanic lakes, and from the mid-ocean ridge system. They estimate that there are about 400 volcanoes active today. Active eruptions only cause around 2 million tons CO_2 per year: Deep Carbon Observatory (2019), "Scientists quantify global volcanic CO_2 venting; estimate total carbon on Earth." EurekAlert, 1 October, https://tinyurl.com/deep-carbon-obs.

This comes close to the estimate of 300 million tons CO_2 by the British Geological Survey that I cited in the last edition of this book and is higher than the estimate of 130–230 million tons CO_2 by the US Geological Survey: Hards, V. (2005), "Volcanic contributions to the global carbon cycle." *British Geological Survey Occasional Publication No. 10* https://tinyurl.com/hards2005; US Geological Survey (2016), "Understanding volcanic hazards can save lives," https://tinyurl.com/usgs2016.

3 The sulfur oxide, ash, and other particles released by the eruption of Mount Pinatubo is thought to have led to a net cooling of 0.5°C: *Scientific American* (2009), "Are volcanoes or humans harder on the atmosphere?," 11 February, https://tinyurl.com/sci-amer-volcanoes.

4 In their 2019 report, FIFA estimates that the footprint of the 2010 World Cup was 41,400 tons CO_2e but excludes the emissions from venues (including stadium construction and energy use in venues) and transport of fans. A 2009 report that did include these came to 2,753,000 tons instead, so I went for their figure for the 2010 World Cup. In their estimates for the 2014 and 2018 World Cups, FIFA included these important aspects, as well as merchandise production. If you've read the book from the start, you will have gathered already that this list is just the easy bits, and you could happily double the footprint if you were a bit more inclusive. It's best not to get too bothered on this occasion: Department of

Environmental Affairs and Tourism (Republic of South Africa) and the Norwegian Government (NORAD) (2009), "Feasibility study for a carbon neutral 2010 FIFA World Cup in South Africa," https://tinyurl.com/fifa-2010; FIFA (2019), "Carbon management and climate protection at FIFA," https://tinyurl.com/fifa-2019.

5 Statista (2020), "Average and total attendance at FIFA football World Cup games 1930–2018," https://tinyurl.com/statista-fifa, based on Sport.de (2018), "WM-Zuschauer: Russland fällt gegenüber Brasilien ab," 15 July, https://tinyurl.com/sport-fifa.

6 FIFA estimates that 3.57 billion people watched at least one minute remotely, including live, delayed, or repeated screenings and highlights. This includes 3.262 billion TV viewers and 309.7 million viewers online or in public viewings. In total, it comes to 34.7 billion viewer-hours: FIFA, "Global broadcast and audience summary," https://tinyurl.com/fifa-audience.

7 Willis, R., Berners-Lee, M., Watson, R., and Elm, M. (2020), "The case against new coal mines in the UK." Green Alliance, January, https://bit.ly/37fJTFU.

8 Digiconomist estimates that, in 2019, Bitcoin's electricity demand was 73.1 TWh, which comes in at 46 million tons of CO_2e, assuming a global electricity mix with a carbon intensity of 0.63 million tons CO_2e per TWh: Digiconomist (2020), "Bitcoin energy consumption index," https://tinyurl.com/digiconomist-bitcoin.

Bitcoin had a market capitalization of 68 percent in January 2020, https://coinmarketcap.com/charts/. So if we assume that other cryptocurrencies have the same carbon footprint per dollar value, the carbon footprint of all cryptocurrencies would be around 68 million tons CO_2e.

9 This is out of 55.6 billion tons CO_2e GHG emissions, including land use change, in 2018: Olivier, J.G.J., and Peters, J.A.H.W. (2019), "Trends in global CO_2 and total greenhouse gas emissions: Summary of the 2019 report," 4 December, PBL Netherlands Environmental Assessment Agency, The Hague, https://bit.ly/2W8wuLJ (see *The world's annual emissions*, p. 194).

10 According to the International Energy Agency (IEA), global electricity demand was more than 23,000 TWh in 2018: IEA (2019), "Global energy & CO_2 status report 2019," https://tinyurl.com/global-electricity.

My figure of 0.32 percent for Bitcoin's share of global electricity is very close to Digiconomist's estimate of 0.33 percent: Digiconomist (2020), see note 8, above.

11 Krause and Tolaymat calculated that the amount of electricity required to mine a single coin increased from 1,074 kWh in January 2016, to 4,577 kWh in January 2017, and 23,157 kWh in January 2018: Krause, M.J., and Tolaymat, T. (2018), "Quantification of energy and carbon costs for mining cryptocurrencies." *Nature Sustainability*, 1 (11), 711–718, https://tinyurl.com/

krause2018. Applying a global electricity mix carbon intensity of 0.63 kg CO_2e per kWh, that is 677 kg CO_2e, 2,884 kg CO_2e, and 14,559 kg CO_2e per coin in 2016, 2017, and 2018, respectively.

12 Bendiksen, C., and Gibbons, S. (2019), "The Bitcoin mining network—trends, average creation cost, electricity consumption & sources," CoinShares Research, https://tinyurl.com/coinshares2019. This report estimates that 73 percent of Bitcoin mining is powered by renewable energy. It also claims that 65 percent of global mining happens in China, and 54 percent in the Chinese province Sichuan. However, China has a higher carbon intensity of electricity than any other developed country and, even though Sichuan has hydropower facilities, the energy that can be derived from hydropower is highly seasonal, so that alternative energy sources such as coal are required: de Vries, A. (2019), "Renewable energy will not solve Bitcoin's sustainability problem," *Joule*, 3(4), 893–898, https://tinyurl.com/deVries2019.

13 Mora et al. argue that, if Bitcoin is taken up similarly to other popular technologies, it could emit almost 23.7 billion tons of CO_2e between 2017 and 2030, and 1.2 billion tons in 2030 alone: Mora, C., Rollins, R.L., Taladay, K., Kantar, M.B., Chock, M.K., et al. (2018), "Bitcoin emissions alone could push global warming above 2°C." *Nature Climate Change*, 8(11), 931–933, https://tinyurl.com/bitcoin-predictions. However, their methodology and assumptions have been questioned by Masanet et al.: Masanet, E., Shehabi, A., Lei, N., Vranken, H., Koomey, J., and Malmodin, J. (2019), "Implausible projections overestimate near-term Bitcoin CO_2 emissions." *Nature Climate Change*, 9(9), 653–654, https://tinyurl.com/masanet2019.

14 Estimates of data centers' use-phase electricity vary between 200 and 300 TWh. At the lower end, the International Energy Agency (2017) estimates that data centers worldwide used 200 TWh in 2020. At the higher end, Andrae (2019) estimates 299 TWh—that would be 126–188 million tons CO_2e, based on a global average electricity of 0.63 million tons CO_2e per TWh. There are also higher estimates out there, such as Belkhir and Elmeligi's (2018) 495 million tons in 2020, but Belkhir himself admits that this figure was somewhat overestimated. These figures are based on older data: IEA (2017), "Digitalisation and energy," https://tinyurl.com/iea2017; Andrae, A.S. (2019), "Comparison of several simplistic high-level approaches for estimating the global energy and electricity use of ICT networks and data centers." *International Journal*, 5, 51, https://tinyurl.com/andrae2019; Belkhir, L., and Elmeligi, A. (2018), "Assessing ICT global emissions footprint: Trends to 2040 & recommendations." *Journal of Cleaner Production*, 177, 448–463, https://tinyurl.com/belkhir2018.

The most recent data come from Masanet et al. and Malmodin, who estimated that data centers consumed 205 TWh in 2018, equal to 129 million tons CO_2e assuming global average electricity: Masanet, E., Shehabi, A.,

Lei, N., Smith, S., and Koomey, J. (2020), "Recalibrating global data center energy-use estimates." *Science*, 367(6481), 984–986, https://tinyurl.com/masanet2020.

In 2020, Jens Malmodin estimated data centers use 230 TWh in 2020, equal to 145 million tons CO_2e based on global average electricity (personal communication; this is an update to an earlier study from 2018): Malmodin, J., and Lundén, D. (2018), "The energy and carbon footprint of the global ICT and E&M sectors 2010–2015." *Sustainability*, 10(9), 3027, https://tinyurl.com/Malmodin2018.

These figures include electricity used to run and cool the servers and any backup power supplies and operational overheads, but not emissions embodied in the servers and the building, although they probably only add a small share relative to the use-phase emissions. For 2015, Malmodin and Lundén estimated embodied emissions of data centers at 9 million tons CO_2e or 15 million tons CO_2e when adjusted for truncation error. That's 10 percent of total data center emissions. If that ratio is applied to 2020, I get to an embodied footprint of 16 million tons CO_2e. Together with the use-phase footprint, that's 161 million tons CO_2e. This is a very rough estimate.

I assume a global average electricity of 0.63 million tons CO_2e per TWh here. Some data center providers buy or generate their own renewable energy, but we don't know the exact share of renewables. Most of the world's data centers are located in the US and the second most in the Asia-Pacific region (by number of servers and workload), according to the IEA: IEA (2017), "Digitalisation and energy," https://tinyurl.com/iea2017.

Both areas have higher electricity carbon intensity than the world average. Greenpeace (2019) points out that 73 percent of China's data centers are powered by coal (https://tinyurl.com/greenpeace2019), that their energy consumption is expected to go up, and that areas where data centers concentrate have one of the lowest shares of renewable energy.

Since 2010, Greenpeace, through their #ClickClean campaign, has been calling on major internet companies to power their data centers on renewable energy. Several large organizations, such as Google and Facebook, have committed to this initiative. Check out the renewable energy scores of your favorite apps at the following link: https://tinyurl.com/clickclean2017.

15 According to the International Energy Agency, the global electricity demand was more than 23,000 TWh in 2018: IEA (2019), "Global energy & CO_2 status report 2019," https://tinyurl.com/global-electricity. The 1 percent of global electricity is also backed up by Masanet et al.: Masanet et al. (2020), see note 14, above. Global GHG emissions were 55.6 billion tons CO_2e, including land use change, in 2018, as detailed in PBL Netherlands Environmental Assessment Agency, The Hague: Olivier and Peters (2019), see note 9, above.

16 It is uncertain what exactly will happen with data centers' footprint in the future. The demand for more data center service capacity is continuing to increase. At the same time, efficiency improvements in processor technology are slowing down and Moore's law is coming to an end: Waldrop, M.M. (2016), "The chips are down for Moore's law." *Nature News*, 530(7589), 144, https://tinyurl.com/waldrop2016. Whether efficiency gains, such as those following Moore's law, automatically reduce world emissions, as many assume, or end up stimulating higher emissions is an important and controversial point.

17 This is an example of Jevons paradox (see p. 188), where efficiency improvements lead to increased resource use because of rebound effects, which is explored in Berners-Lee, M., and Clark, D. (2013), *The Burning Question: We Can't Burn Half the World's Oil, Coal and Gas. So How Do We Quit?* (London: Profile Books).

There is also this report by the UK Energy Research Centre: Sorrell, S. (2007), "The rebound effect: An assessment of the evidence for economy-wide energy savings from improved energy efficiency," https://tinyurl.com/sorrell2007.

18 The two top figures are provisional data for 2019 for territorial CO_2 and greenhouse gas emissions from the UK's Office for National Statistics: ONS (2020), "Provisional UK greenhouse gas emissions national statistics," https://tinyurl.com/ons-uk-emissions. The third figure is based on my own modeling, using my input-output model.

19 The figure for the USA comes from reporting from the Environmental Protection Agency (EPA), https://tinyurl.com/EPAUSAghgemissions.

20 The data are based on: Ritchie, H., and Roser, M. (2019), "CO_2 and greenhouse gas emissions: Consumption-based (trade-adjusted) CO_2 emissions." Our World in Data, https://tinyurl.com/emissions-by-country.

21 Based on data from: Ritchie, H., and Roser, M. (2019), "CO_2 and greenhouse gas emissions: Production vs consumption-based CO_2 emissions per capita." Our World in Data, https://tinyurl.com/emissions-per-capita.

22 Tim Jackson's classic book *Prosperity Without Growth*, updated in 2017 (Abingdon: Routledge), is a rigorous and accessible articulation of this uncomfortable reality. You can read the original 2009 article for free here: https://timjackson.org.uk/ecological-economics/pwg/.

I also recommend Kate Raworth's book *Doughnut Economics: Seven Ways to Think Like a 21st-Century Economist*, published in 2017 (London: Business Books/Penguin), and her 2018 TED Talk "A healthy economy should be designed to thrive, not grow," https://tinyurl.com/raworth-tedtalk.

23 Based on consumption-based emissions from Our World in Data, divided by total GDP of each country provided by the World Bank: https://tinyurl.com/world-bank-gdp.

Billions of tons

1 The data used here were calculated from the Global Fire Emissions Database: www.globalfiredata.org/.

2 Paddison, L. (2019) "2019 was the year the world burned," *Huffington Post*, 27 December, https://tinyurl.com/paddison2019, https://bit.ly/2Vpl8m4.

3 *Nature* (2020), "Playing with fire could turn the Amazon into a carbon source," 10 January, https://go.nature.com/34EAAPe; Brando, P.M., Soares-Filho, B., Rodrigues, L., Assunção, A., Morton, D., Tuchschneider, D., et al. (2020), "The gathering firestorm in southern Amazonia." *Science Advances*, 6(2), eaay1632, https://tinyurl.com/brando2020.

4 In the end (with a lot of help from colleagues), I found that the three most credible studies roughly converged on the same ballpark figure, once they had been adjusted to take account of the different things they excluded and included.

Malmodin and Lundén estimated the carbon footprint of the IT industry (including data centers, networks, and user devices like computers, smartphones, and traditional phones) at 730 million tons CO_2e and an additional 420 million tons CO_2e for TVs, TV networks, and other consumer electronics (such as cameras, projectors, non-smart speakers, and portable media players like iPods and game consoles), coming to a total of 1,150 million tons CO_2e. In conversation, Jens Malmodin provided me with his most recent estimates for 2020: 690 million tons CO_2e for data centers, networks, and user devices in 2020 and a further 400 million tons CO_2e for TVs, TV networks, and other consumer electronics, coming to a total of 1.1 billion tons CO_2e: Malmodin, J., and Lundén, D. (2018), "The energy and carbon footprint of the global ICT and E&M sectors 2010–2015." *Sustainability*, 10(9), 3027, https://tinyurl.com/Malmodin2018.

Andrae also arrives at 1.1 billion tons CO_2e for the IT industry including TV: Andrae, A.S. (2019), "Comparison of several simplistic high-level approaches for estimating the global energy and electricity use of ICT networks and data centers." *International Journal*, 5, 51, https://tinyurl.com/andrae2019.

Belkhir and Elmeligi's estimates are a bit higher, at between 1.1 and 1.3 billion tons CO_2e for ICT without TVs: Belkhir, L., and Elmeligi, A. (2018), "Assessing ICT global emissions footprint: Trends to 2040 & recommendations." *Journal of Cleaner Production*, 177, 448–463, https://tinyurl.com/belkhir2018.

All these estimates include use-phase and embodied emissions, but their methodologies incur truncation error (the omission by life-cycle analyses of large numbers of small supply chain pathways that are collectively very significant). When this is adjusted for, Malmodin's (2018) estimate of the IT's footprint rises to 870 million tons CO_2e for data centers, networks, and user devices, and 460 million tons CO_2e for TVs, TV equipment like set-top boxes, TV networks, and other consumer electronics, for a total of 1.3 billion tons CO_2e. Andrae's (2019) estimate rises to 1.4 billion tons CO_2e, and Belkhir and Elmeligi's (2018) estimates rise to 1.5–1.8 billion tons CO_2e, or 1.9–2.2 billion tons CO_2e if Malmodin's figure for TVs, TV networks, and consumer electronics is included in their estimate.

I will go with Malmodin's estimate plus cryptocurrencies (see *Cryptocurrencies*, p. 178), coming to a total of 1.4 billion tons CO_2e. These estimates include some, but not most, embedded IT in "smart" devices, including the "Internet of Things."

5 Malmodin and Lundén (2018; see note 4, above) estimate that 44 percent of user devices' footprint in 2015 came from embodied emissions; Belkhir and Elmeligi (2018; see note 4, above) estimate between 62 percent and 64 percent in 2020. This is particularly pronounced for smartphones, where 85–95 percent of the footprint is from embodied emissions: Belkhir, L. (2018), "How smartphones are heating up the planet." *The Conversation*, 25 March, https://tinyurl.com/belkhir-smartphones.

An exception to this rule is set-top boxes and routers, because most people leave them on 24/7. My own analysis suggests that most of the footprint of computers (p. 140), smartphones (p. 126), and TVs (p. 47) is embodied.

6 William Stanley Jevons (1865), *The Coal Question: An Inquiry Concerning the Progress of the Nation, and the Probable Exhaustion of Our Coal-Mines* (London and Cambridge: Macmillan). Jevons pointed out that more efficiency would make coal more attractive and would increase demand rather than reduce it. I have written about this in *There Is No Planet B* (2019), which is in many ways a companion to *Bananas*, and before that, with Duncan Clark, even more extensively, in *The Burning Question* (2013).

7 The atomic bomb dropped on Hiroshima in 1945 had an explosive energy of 15 kilotons of TNT equivalent: Atomic Heritage Foundation (2014), "Little Boy and Fat Man," https://tinyurl.com/hiroshima-bomb.

8 Based on a conversion factor of \$1 = £0.81 at the time of writing. I'm using estimates of total spend and multiplying these by the carbon intensity per £1 spent of the UK's defense sector of 0.235 kg CO_2e per £1 based on my input-output model. However, I multiply it by 1.5 for the US and 2.3 for the world because their economy is that much more carbon intensive in terms of CO_2 per \$ GDP (see p. 184 for the UK's and US's carbon intensity per \$). For the world, this is based on 35.7 million tons CO_2 in 2016 according to Our World in Data (https://tinyurl.com/annual-co2) and 76 trillion \$ GDP based on the World Bank (https://tinyurl.com/world-bank-gdp).

9 Neta Crawford, co-director of the The Costs of War Project by the Watson Institute at Brown University, estimates US spending on the

Iraq War at $1,922 billion: Crawford, N.C. (2020), "The Iraq War has cost the US nearly $2 trillion." *Defense One*, 4 February, https://tinyurl.com/crawford-2020.

10 The Stockholm International Peace Research Institute estimates that the US spent $730 billion in 2019 on defense, while the world total spend on defense was $1,910 billion. The UK spend is estimated at $49 billion: Tian, N., Kuimova, A., Da Silva, D.L., Wezeman, P.D., and Wezeman, S.T. (2020), "Trends in world military expenditure, 2019." Stockholm International Peace Research Institute, https://tinyurl.com/tian2020.

11 The Costs of War Project by the Watson Institute at Brown University has estimated the US military's footprint for the fiscal years 2001–2018 and concluded that the US Department of Defense is "the world's largest institutional user of petroleum and correspondingly, the single largest institutional producer of greenhouse gases (GHG) in the world." Crawford, N. (2019), "Pentagon fuel use, climate change, and the costs of war." Watson Institute for International & Public Affairs, Brown University, https://tinyurl.com/crawford-2019.

12 There is a worrying trend in the Syrian civil war for attacks to focus on hospitals. Not only does this increase suffering and defy international humanitarian law, but it also means that the destroyed facilities need to be rebuilt, which comes at a carbon cost: Koteiche, R. (2019), "Destroying hospitals to win the war." Physicians for Human Rights, 21 May, https://tinyurl.com/koteiche2019.

13 Stuart Parkinson from Scientists for Global Responsibility estimates that the global military footprint might be 5 percent of global emissions and another 1 percent for war impacts in 2018. Of the 55.6 billion tons CO_2e emitted in 2018, that's 3.3 billion tons: Parkinson, S. (2019), "The carbon bootprint of the military," 29 June, https://tinyurl.com/parkinson-2019.

14 Jacobson, M.Z. (2009), "Review of solutions to global warming, air pollution, and energy security." *Energy & Environmental Science*, 2(2), 148–173, https://tinyurl.com/jacobson2009.

15 This article in *Nature* summarizes several recent studies on this topic: Witze, A. (2020), "How a small nuclear war would transform the entire planet." *Nature*, 579(7800), 485–487, https://tinyurl.com/witze202.

The figure for a regional nuclear war comes from: Toon, O.B., Bardeen, C.G., Robock, A., Xia, L., Kristensen, H., McKinzie, M., et al. (2019), "Rapidly expanding nuclear arsenals in Pakistan and India portend regional and global catastrophe." *Science Advances*, 5(10), eaay5478, https://tinyurl.com/toon2019.

A regional conflict could cause 16–36 million tons of black carbon to be released and lead to a cooling of the climate by 2°C to 5°C globally. There would also be 50 to 125 million deaths.

The estimate for a nuclear war between Russia and the US comes from: Robock, A., Oman, L., and Stenchikov, G.L. (2007), "Nuclear winter revisited with a modern climate model and current nuclear arsenals: Still catastrophic consequences." *Journal of Geophysical Research: Atmospheres*, 112(D13), https://tinyurl.com/robock2007. A US-Russia nuclear war could release 150 million tons of black carbon and lead to a 10°C drop in global temperatures.

16 Klare, M.T. (2020), "How rising temperatures increase the likelihood of nuclear war." *The Nation*, 13 January, https://tinyurl.com/klare2020. The effect of climate change on agriculture, competition for the polar waters, and the frequency of natural disasters could fuel international conflicts and increase the risk of a nuclear war.

17 Watson, J.E., Evans, T., Venter, O., Williams, B., Tulloch, A., Stewart, C., et al. (2018), "The exceptional value of intact forest ecosystems." *Nature Ecology & Evolution*, 2(4), 599–610, https://tinyurl.com/watson2018-forests.

18 UN REDD Programme (2019), "Forest facts," www.un-redd.org/forest-facts.

19 Curtis, P.G., Slay, C.M., Harris, N.L., Tyukavina, A., and Hansen, M.C. (2018), "Classifying drivers of global forest loss." *Science*, 361(6407), 1108–1111, https://tinyurl.com/curtis-2018.

20 Woodland Carbon Code (2018), "Woodland carbon calculation spreadsheet," https://tinyurl.com/woodland-carbon.

21 Gov.uk Guidance (2019), "Woodland Carbon Guarantee," https://tinyurl.com/woodland-guarantee and One Tree Planted, a global initiative with projects and events in the US, https://onetreeplanted.org/collections/united-states.

22 Bond, T.C., Doherty, S.J., Fahey, D.W., Forster, P.M., Berntsen, T., DeAngelo, B.J., et al. (2013), "Bounding the role of black carbon in the climate system: A scientific assessment." *Journal of Geophysical Research: Atmospheres*, 118(11), 5380–5552, https://tinyurl.com/bond-2013.

23 Bond et al. (2013). See note 22, above.

24 Timonen, H., Karjalainen, P., Aalto, P., Saarikoski, S., Mylläri, F., Karvosenoja, N., et al. (2019), "Adaptation of black carbon footprint concept would accelerate mitigation of global warming." *Environmental Science & Technology*, 53, 12153–12155, https://tinyurl.com/timonen2019.

25 Timonen et al. (2019). See note 24, above.

26 Lund, M.T., Samset, B.H., Skeie, R.B., Watson-Parris, D., Katich, J.M., et al. (2018), "Short black carbon lifetime inferred from a global set of aircraft observations." *npj Climate and Atmospheric Science*, 1(1), 1–8, https://tinyurl.com/lund-2018.

27 Bond et al. (2013). See note 22, above.

There are other estimates of the mass of black carbon released every year and its global warming potential, such as by the Climate and Clean Air Coalition (https://tinyurl.com/ccacoalition), which estimates emissions at 6.6 million tons in 2015 and cites a global warming potential of 460–1,500. But on average they also arrive at 6.5 billion tons CO_2e, not too far off the estimate of 8.8 billion.

28 Data for 2018 from Olivier and Peters and the Global Carbon Project. I've used a markup factor of 1.9 for high-altitude emissions, as suggested by the UK government's Department for Business, Energy and Industrial Strategy (BEIS), and approximated aviation emissions at 2 percent of CO_2 without that: Olivier, J.G.J, and Peters, J.A.H.W. (2019), "Trends in global CO_2 and total greenhouse gas emissions: Summary of the 2019 report," 4 December, PBL Netherlands Environmental Assessment Agency, The Hague, https://bit.ly/2W8wuLJ; Global Carbon Project (2019), https://tinyurl.com/carbon-project.

29 The UN's *Emissions Gap Report* 2019 concluded that "There is no sign of GHG emissions peaking in the next few years; every year of postponed peaking means that deeper and faster cuts will be required." UN Environment Programme (2019), *Emissions Gap Report* 2019. UNEP, https://tinyurl.com/emission-gap.

30 Nobody knows exactly how much fuel there is in the ground, so these numbers are estimates. The proven reserves come from BP's Statistical Review of World Energy 2019, https://tinyurl.com/bp-review2019.

The figures for unconventional fuels like tar sands come from: Center for Sustainable Systems (2019), "Unconventional fossil fuels factsheet." University of Michigan, https://tinyurl.com/unconv-fossil.

Up-to-date data on recoverable resources was hard to come by, so I have fallen back on an old source: International Energy Agency (IEA) (2012), "World energy outlook 2012," https://tinyurl.com/iea-2012.

This is the same source that Duncan Clark and I used in *The Burning Question* (2013), which covers the abundance of fossil fuel in a lot more detail.

Negative emissions

1 Bastin, J.F., Finegold, Y., Garcia, C., Mollicone, D., Rezende, M., Routh, D., et al. (2019), "The global tree restoration potential." *Science*, 365(6448), 76–79, https://tinyurl.com/bastin2019.

2 Veldman, J.W., Aleman, J.C., Alvarado, S.T., Anderson, T.M., Archibald, S., Bond, W.J., et al. (2019), "Comment on 'The global tree restoration potential.'" *Science*, 366(6463), eaay7976, https://tinyurl.com/veldman2019.

3 Oreska, M.P., McGlathery, K.J., Aoki, L.R., Berger, A.C., Berg, P., and Mullins, L. (2020), "The greenhouse gas offset potential from seagrass restoration." *Scientific Reports*, 10(1), 1–15, https://tinyurl.com/oreska2020.

4 This is an estimate based on data from Duarte et al. (2013), which estimates that current global carbon storage for marine vegetation is 19.65 billion tons (3.45, 9.9, and 6.3 billion tons of carbon storage for salt marshes,

mangroves, and seagrasses, respectively) and roughly 35 percent losses in marine vegetation since the World War II (25 percent for salt marshes, 40 percent for mangroves, and 30 percent for seagrasses). This adds up to around 10 billion tons of carbon lost, or 37 billion tons CO_2e: Duarte, C.M., Losada, I.J., Hendriks, I.E., Mazarrasa, I., and Marbà, N. (2013), "The role of coastal plant communities for climate change mitigation and adaptation." *Nature Climate Change*, 3(11), 961–968, https://tinyurl.com/duarte2013.

5 Sanderman, J., Hengl, T., and Fiske, G. J. (2017), "Soil carbon debt of 12,000 years of human land use." *Proceedings of the National Academy of Sciences*, 114(36), 9575–9580, https://tinyurl.com/sanderman2017.

6 Amundson, R., and Biardeau, L. (2018), "Opinion: Soil carbon sequestration is an elusive climate mitigation tool." *Proceedings of the National Academy of Sciences*, 115(46), 11,652–11,656, https://tinyurl.com/amundson2018.

7 Brownsport, P., Carter, S., Cook, J., Cunningham, C., Gaunt, J., Hammond, J., et al. (2010), "An assessment of the benefits and issues associated with the application of biochar to soil." Defra, https://tinyurl.com/shackley2010.

8 Paustian, K., Lehmann, J., Ogle, S., Reay, D., Robertson, G.P., and Smith, P. (2016), "Climate-smart soils." *Nature*, 532(7597), 49–57, https://tinyurl.com/paustian2016.

9 Psarras, P., Krutka, H., Fajardy, M., Zhang, Z., Liguori, S., Dowell, N.M., and Wilcox, J. (2017), "Slicing the pie: How big could carbon dioxide removal be?" *Wiley Interdisciplinary Reviews: Energy and Environment*, 6(5), e253, https://tinyurl.com/psarras2017.

10 Moosdorf, N., Renforth, P., and Hartmann, J. (2014), "Carbon dioxide efficiency of terrestrial enhanced weathering." *Environmental Science & Technology*, 48(9), 4809–4816, https://tinyurl.com/moosdorf2014, doi:10.1088/1748-9326/aaa9c4; and Strefler, J., Amann, T., Bauer, N., Kriegler, E., and Hartmann, J. (2018), "Potential and costs of carbon dioxide removal by enhanced weathering of rocks." *Environmental Research Letters*, 13(3), 034010, https://tinyurl.com/strefler2018.

11 Strefler et al. (2018). See note 10, above. Thanks for this to Chris Goodall and his superb weekly newsletter *Carbon Commentary* (an excellent source of low-carbon technology updates with free subscription), www.carboncommentary.com/.

12 Psarras et al. (2017). See note 9, above.

What can we do?

1 For the average US person's carbon footprint, I have used my environmentally extended input-output (EEIO) model to get a final number of 21 tons CO_2e. This is roughly in line with a recent UN report which estimated an average footprint of 17.5 tons of CO_2 only (so adding on an extra 20 percent

for non-CO_2 greenhouse gas emissions would take us up to just over 21 tons).

2 Wolff, E.N. (2017), "Household wealth trends in the United States, 1962 to 2016: Has middle class wealth recovered?" National Bureau of Economic Research, working paper 24085, https://www.nber.org/system/files/working_papers/w24085/w24085.pdf.

3 For the pie chart of the average US person's carbon footprint I have drawn on several different sources.

First of all, for the food: the average US person consumes around 3,700 calories per day (according to Our World in Data, https://ourworldindata.org/food-supply), which is almost twice as much as needed. The average US person also wastes a massive 32 percent of their food (according to Yu and Jaenicke (2020), "Estimating food waste as household production inefficiency," *American Journal of Agricultural Economics*, 102(2), 525–547, https://preview.tinyurl.com/usfoodwaste), and has a high amount of meat and dairy in their diet. This equals around 81 kg CO_2e per person per week or 4.2 tons of CO_2e per person per year on the food they buy, plus another 0.4 tons CO_2e per year for eating out.

Next, home and accommodation: the average US household consumes a massive 10,600 kWh per year on electricity and another 11,700 kWh per year on gas (according to the US Energy Information Administration, https://tinyurl.com/ushomeenergy). This amounts to 3.5 tons CO_2e on electricity and 1.4 tons CO_2e for gas per person per year. I have estimated that the house itself adds around 1 ton per person per year and staying away from home adds another 0.5 tons.

For travel: the average US car covers around 13,500 miles per year and averages 0.63 kg CO_2e per mile. Based on an average of three passengers per car, this comes out to 2.82 tons. I have also estimated that the car itself adds on another 0.94 tons per year. The average US person flies around 1,500 miles per year, equating to a further 0.55 tons CO_2e. Another 0.2 tons goes on ferries and cruises and a further 0.28 tons on trains, buses, and public transit (these numbers I have lifted from the average UK person's annual footprint. While Americans use public transport a lot less than Britons, the carbon intensity of public transport tends to be higher, so I think they are unlikely to be significantly different in the US. And in any case, it's such a small component of the overall footprint that it doesn't really make much difference).

Finally, there is everything else: the figures for bought services, water waste, and sewage; leisure, recreation, and attractions; and non-food shopping I have kept the same as the average UK person, again because they are unlikely to be significantly different. For health, education, and other public services, I have applied a three-times markup factor compared to the average UK person (see *An operation*, p. 146). The reason for this is that the US has a healthcare footprint that is three times larger per person compared to the average UK person and a military footprint that is a massive four-and-a-half times as large as the average UK person (see *A war*, p. 190). I haven't got a number for education and other public services, but I believe this will be more in line with what we have in the UK. Based on all this, I have applied a markup of factor of three compared to the average UK person to arrive at a footprint of 3.8 tons per person per year in the US.

Thanks

My biggest thanks go to Liz, Bill, and Rosie for their brilliant support and understanding, through both the first edition and this update, which turned out to be a far bigger project than I had imagined.

Many thanks to Jess Moss at Small World Consulting for unearthing quirky data and sorting out dozens of references—and for reminding me to get a move on. David Howard, Kim Kaivanto, Andy Scott, and Geraint Johnes from Lancaster University and Sonny Khan all helped with the input-output model that I have drawn upon extensively. Thanks also to David Parkinson and Chris Goodall, among others, for answering technical queries.

Andrew Meikle let me chatter away while ride-sharing and has been a frequent sounding board. He read early pages aloud so I could hear how bad they were. Others who cast a friendly eye include Phil and Jane Latham, Aly Purcell, Rachel Nunn, and Mark Jameson. Mum and Dad, true to form as incredible parents, both picked through the entire draft at a moment's notice.

The original (2010) edition could not have happened without Duncan Clark, whose edits and advice were superb throughout the project. Or Kim Quazi, who helped me thrash out the first ideas years ago in a pub. Going even further back, the list of people to be grateful to is clearly endless, but I want to thank Lee Pascal, David Brazier, and Simon Loveday for three very different contributions.

I'm grateful to many of Small World's clients for providing material, but especially to Booths Supermarkets, Lancaster University, the Crichton Carbon Centre, Historic Scotland, and Everards Brewery.

For this 2020 update, I had an army of helpers, including Sam Allan, Rosie Berners-Lee, Alex Boyd, Tom Davies (who pulled out all the stops to get the first draft over the line), Sarah Donaldson, Charlie Freitag (who pulled out more stops), Matt Jones, Katerina Karpasitou, Cordelia Lang, Lou Mamalis, Tom Mayo, Helena Pribyl, Lorraine Ritchen-Stones, and Rosie Watson. It was brilliant to work with Nick Hewitt on many of the papers that have fed into this book and thanks also to Robin Frost for years collaborating on models.

A ton of work went into unscrambling the carbon footprint of the digital world and some top academics were generous with their time. Thanks especially to Anders Andrae, Lotfi Belkhir, Livia Cabernard, Jens Malmodin, Chris Preist, Paul Shabajee, and Daniel Schien, as well as the team from Lancaster University; and to Gordon Blair, Adrian Friday, Bran Knowles, and Kelly Widdicks, who worked with Charlie Freitag and me on a project for the Royal Society. Thanks to Jigna Chandaria and the BBC for giving us access to TV analysis.

At Profile, huge thanks to Mark Ellingham who has been a patient and skilled editor. Also to Henry Iles for page layout and straightening out diagrams, Bill Johncocks for the index, and Nicky Twyman for the major task of proofreading.

For this North American edition, yet more thanks are due to Tom Davies who did a fine job working through the myriad of careful alterations needed make this book work just as well on the other side of the Atlantic from its birthplace in the UK. (When you think about it, everything is a bit different: cars, foods, lifestyles, geography, units of measure, language—the lot!)

Lastly, thanks for all the emails over the years with encouragement, feedback, and stories of bananas instead of roses for Valentine's Day. Please keep writing to info@howbadarebananas.com.

Index

NOTE: The index covers the main text and the Appendix but not the Notes and references. Bold page numbers indicate that relevant information on that page is in a diagram or table.